Birds of Clar
1982

A Ten Year Report

Compiled and edited by
Liam Lysaght, Tony Mee, John Murphy and Tom Tarpey

Published jointly by:
the Clare and Limerick branches of the Irish
Wildbird Conservancy,
Ballyorgan,
Kilfinane,
Co. Limerick

March 1994

Copies of this report are available from:
Tony Mee at the above address.

ISBN 0 9523348 0 1

Printed by: Galway City Print,
                    Prospect Hill,
                    Galway.

Drawings copyright the artists

Front cover by Phil Brennan

Acknowledgement
Thanks to Aughinish Alumina, Limited &
Electricity Supply Board, Limerick
for a contribution to defray some of the publication
costs.

Contents

Irish Wildbird Conservancy

The Irish Wildbird Conservancy (IWC) is the largest independent conservation organisation in Ireland, with a nationwide network of local branches and a membership of over 4000. The IWC is a registered charity and is financed by membership subscriptions and donations.

We are devoted to all matters pertaining to wild birds and their environment, including conservation, education and research. The IWC owns and manages a number of nature reserves and undertakes frequent national bird surveys. We are the Irish national partner of *Birdlife International*.

A small professional staff is based at our headquarters at Ruttledge House, 8 Longford Place, Monkstown, County Dublin. Tel. 01-2804322 or 01-2808237, Fax. 01-2844407.

There are local branches in Clare and Limerick. New members are always welcome and you do not have to be an expert to join! For information and membership forms please contact Irish Wildbird Conservancy headquarters.

Both branches work on different conservation projects in their county and also organise regular events for the public, i.e lectures, slide shows and outings to popular bird watching sites. These events are open to the general public and are advertised well in advance. A list of events is also available from the above address.

Editorial

This is the third report to cover both counties Clare and Limerick. The previous report *Birds of North Munster* (Brennan and Jones 1982) covered the years 1972 to 1981 and included the county of Tipperary and part of north Kerry in its area of coverage. This report strives to give as full an update as possible for the intervening ten years 1982 tó 1991, while confining itself to the counties of Clare and Limerick. However, all records received from sites which straddle the border of a neighbouring county, eg. Charleville Lagoons which is partially in county Cork, have been included in this report.

The years 1982 to 1991 have marked an era of considerable change within the region on a number of fronts. Following on the pioneering work of Ewart Jones and Philip Buckley at Loop Head and the Bridges of Ross in the late seventies and early eighties, the intervening years have seen the Bridges of Ross firmly established as the premier 'seawatch' point on Ireland's west coast, attracting an ever increasing number of birders into the region from elsewhere in Ireland and abroad. Since 1985, attention has also focussed on nearby Kilbaha and Loop Head as an autumn land fall area for migrants, particularly passerines. Spearheaded by Phil Brennan and based at Kilbaha, this area has been put on 'observatory' footing each autumn with careful logging of all species both resident and migrants. The fruits of this upsurge of interest in the Loop Head peninsula on both land and sea are reflected in the species list and in some of the articles included in this report. A total of forty new species for the region was recorded during the ten years of this report, of which six originate from the Bridges of Ross and twenty-one from Loop Head/Kilbaha. This shows a significant rise when compared to the previous ten years 1972-1981 when twelve new species were recorded in the region.

While 'rarity hunting' has become a feature of the birding scene in the region as elsewhere, local participation in national surveys also played its part. Fieldwork for the *Atlas of Wintering Birds in Britain and Ireland, Winter Wetlands Survey* and *The New Atlas of Breeding Birds of Britain and Ireland: 1988-1991.* kept fieldworkers busy during the period of this report, and actually led to the discovery of some new sites of ornithological interest such as the Charleville Lagoons.

Another major change which occurred during the period was the formation of three new IWC county branches for Clare, Limerick and Tipperary, replacing the old North Munster Branch. It is still early days for both the Clare and Limerick branches but longer-term it should result in a stronger, more vibrant IWC branch network.

Unfortunately some things have not changed. The number of active observers within the two counties remains thin on the ground. A large chunk of this report's species list was contributed by birders from outside the region. On a similar vein, a large number of Clare squares in the *New Atlas of Breeding Birds in Britain and Ireland:1988-1991* were completed by UK fieldworkers employed by the British Trust for Ornithology (BTO) due to insufficient local participation.

We hope that the publication of this report may act as the encouragement needed for a whole new generation of birders interested in furthering our knowledge of the birds of Clare and Limerick.

T.M.

Species List

Introduction

The sequence and scientific nomenclature used in this report follows the method devised by Voous, K.H. (1973-77) in the *List of Recent Holarctic Bird Species*. Species that require substantiating descriptions prior to acceptance by the Irish Rare Birds Committee (IRBC) are marked with an asterix. Records which have not yet been accepted by the IRBC are marked with a double asterisk.

For species that require acceptance by the IRBC the observers name is listed with the record in question. All contributors are also listed in alphabetical order.

Records

Records are very welcome for future reports. They should be submitted as soon as possible but definitly before the 31st January of the following year to the appropriate county recorder:

For Clare:	Phil Brennan	For Limerick:	Tony Mee
	The Crag		Ballyorgan,
	Stonehall,		Kilfinane
	Newmarket-on-Fergus		Co. Limerick
	Co. Clare		

Records submitted should include the following: The species seen; the number seen; date(s) of observation; location and any other relevant comments.

Records of rarities

Rarity descriptions should be send directly to :

Pat Smiddy
IRBC Secretary
Ballykenneally,
Ballymacoda
Co. Cork

A list of species which require a description and standard description forms are also available from Pat.

Contributors

A. Ackerly
R. Ackerly
P. Archer
P. Brennan
P. Buckley
D. Brooks
S. Brooks
S. Busuttil
M. Campbell
T. D. Carruthers
V. Caschera
B.Clarke
J. Clarke
D. Cooke
J.A. Coveney
G. M. Cresswell
A. Cullen
P. Cummins
G. D'Arcy
G. Donnelly
M. Donahue
J.F. Dowdall
S. Enright
M. Evans
K. Fahy
S. Farrell
M.D.E. Fellowes
B. Finnegan

J. Fox
R.A. Frost
K. Grace
J.H. Grant
A.H.J. Harrop
I.D. Halliday
J. Holmes
D.H. Hughes
S.C. Hutchings
C.W. Hutchinson
S.C. Johnson
E.E.H. Jones
A. Kelly
M. Kelly
R. Kiely
T. Kilbane
P. Lonergan
Loop Head Bird Observatory
J.K. Lovatt
L.S. Lysaght
D.G. Mac Adam
E.A. Mac Lochlainn
D. Manley
R. Martins
A. Mee
T. Mee
C. Meehan
M. Meehan

O.J. Merne
C.C. Moore
C. Moriarty
K. Mullarney
C. Murphy
D.F. Murphy
J. Murphy
R. Nairn
M.V. O'Brien
M.J. O'Donahue
M. O'Donnell
J. O'Flaherty
O. O'Sullivan
G. Pearson
S.H. Piotrowiski
K. Preston
C.P.S. Raffles
J. Reynolds
P. Roscoe
V. Sheridan
P. Smiddy
A. D. Smith
T. Tarpey
J. Tottenham
R. Tottenham
Fr. W. Troddyn
N. Walsh
T. Whilde

Systematic list

Red-throated Diver *Gavia stellata*
Passage migrant and scarce winter visitor. Occurs regularly in small numbers at Liscannor Bay and at Ballyvaughan Bay north to Aughinish Island from late October to mid May. An unprecedented count of 84 at Liscannor Bay on 13th January 1983 does not fit into the normal pattern of occurence. Individuals recorded irregularly in winter on the Shannon Estuary from Poulnasherrry Bay to Kilbaha. A single bird, the first for the upper estuary, was recorded at Shannon on 30th March 1991. Recorded regularly on autumn passage at the Bridges of Ross from late September to mid October with a peak autumn total of 44 birds in 1988.

Black-throated Diver* *Gavia arctica*
Passage migrant and winter visitor. The north Clare coastline from Ballyvaughan Bay north to Aughinish Island has become a regular haunt in winter and spring for this species which remains scarce nationally. Records span from early November to late May. Winter concentrations, which are of national importance, peak normally in January and are summarised as follows;

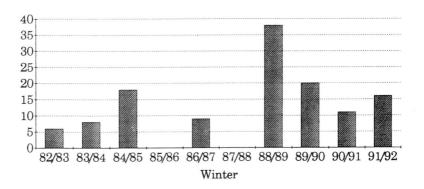

Two at Liscannor Bay on 1st April 1983 is the only record for Clare south of Ballyvaughan Bay. As there are only two records for the region prior to 1982, the recent influx of records is almost certainly due to a greater awareness by observers of the key identification features for this species rather than to any great change in distribution.

Great Northern Diver *Gavia immer*
Regular passage migrant and winter visitor along the Clare coast, with largest concentrations occuring from Black Head north to Aughinish Island. Counts of 146 (storm driven) from the Rine north to Doorus on 27th

January 1990 and 109 from Black Head to Flaggy Shore on 14th February 1990 suggests that the wintering population for this area is greater than previously estimated. There is also a spring build-up off Black Head where a peak of 50 was recorded on 17th April 1982. Summer records from Black Head include three on 5th June 1982 and one on 4th July 1989. Eight at Lahinch on 27th November 1989 is the only significant record from Liscannor Bay. Birds are regularly recorded on autumn passage at the Bridges of Ross, but their passage tends to be later than Red-throated. The occasional storm-blown bird has been recorded in the upper reaches of the Shannon Estuary.

Little Grebe *Tachybaptus ruficollis*
Common resident on most freshwater lakes in both counties though absent from some parts of west Clare. Some local concentrations occur in autumn and winter, with notable peaks of 40 at Ballyallia Lake on 15th January 1991, 37 at Lough Inchiquin on 13th January 1991, 28 at Aughinish, Foynes on 24th September 1982 and 19 at Charleville Lagoons on 22nd August 1987. Also recorded in small numbers in the Shannon Estuary.

Great Crested Grebe *Podiceps cristatus*
Widespread resident and passage migrant, mirroring Little Grebe in terms of distribution but confined more to freshwater lakes that are richer in fringe vegetation. Numbers peak in the Shannon Estuary during August/September with 53 at Clonderlaw Bay on 27th August 1983 being a notable count. Also occurs in good numbers off the north Clare coast in winter.

Red-necked Grebe* *Podiceps grisegena*
Rare winter visitor.
>   1988: One at Ballyvaughan on 20th February (P.Archer, P.Lonergan, et al)
>   1989: One at Ballyvaughan on 19th February (E.A. McLochlainn)

These are the third and fourth records for Clare. There is one previous record for Limerick.

Slavonian Grebe *Podiceps auritus*
Rare winter visitor.
>   1982: Two at New Quay on 4th December, and one at Labasheeda Bay on 20th November
>   1983: Three at New Quay on 22nd January
>   1985: One at Quilty on 13th March
>   1987: One at Ballyvaughan on 1st January.
>   1989: A pair inland at Lough Graney on 24th January
>   1990: One at Ballyvaughan on 29th December

There has been a significant increase in records throughout Clare including one notable inland record. This species has not yet been recorded in Limerick.

Fulmar *Fulmarus glacialis*

Resident and passage migrant. Breeds on suitable cliffs along the west coast of Clare, with by far the largest concentration at the Cliffs of Moher where 3077 adults on nests were counted in June 1987. Sixty-six nest sites were noted at Loop Head on 16th May 1987, but greater concentrations occur on the cliffs north of Kilkee along Donegal Point and Ballard Bay. Both spring and autumn passage occurs at the Bridges of Ross where a maximum of 800 per hour was recorded on 15th August 1990. A count of 850 on 24th January 1982 at the Cliffs of Moher suggests that a significant number of birds may remain to winter close to their breeding cliffs. Occasionally occurs inland after storms with one seen over Shannon Town on 4th September 1983.

'Blue' phase birds have been noted regularly in spring and autumn at the Bridges of Ross with a miminum of 57 recorded from the period 1984 to 1988, including a day total of 12 on 23rd September 1984.

Cory's Shearwater* *Calonectris diomedea*

Rare autumn passage migrant off Loop Head and the Bridges of Ross. Remains very scarce on autumn passage when compared to the south coast.

Records omitted from previous ten year report.
1981: At Loop Head, one on 12th, three on 13th, two on 22nd and 180 on 24th
August (P.Buckley)

All records are from the Bridges of Ross.
1985: Singles on 2nd August and first week of September (P.Buckley)
1988: One on 25th August (A.Cullen)
1990: One on 29th August (K.Mullarney)

Unfortunately the 1981 passage has not been repeated in the intervening years.

Great Shearwater *Puffinus gravis*

Rare autumn passage migrant. All records refer to the Bridges of Ross except where stated:
1984: Three on 9th September, one on 18th September and two on 20th
September
1985: One on 10th August, 11 on 20th August, and two on 28th August.
1988: One on 20th August
1989: One on 12th August
1990: 10 recorded between 15th August and 5th September. Thirteen recorded at
sea on pelagic trip west of Loop Head on 11th August
1991: An extraordinary movement of 1000 during the first hour of daylight on
14th September, six on 16th September and three on 22nd September.

The 1991 movement illustrates that a dawn seawatch can sometimes reap rich dividends.

Sooty Shearwater *Puffinus griseus*
Regular autumn passage migrant. Recorded on passage at the Bridges of Ross from early August to early November with a peak hourly rate of 800 birds on 20th August 1988 and a maximum day total of 3293 on 30th August 1990. A notable feature of the 1986 passage was a feeding raft of *circa* 1200 birds off Loop Head on 1st October increasing to 1650 on 2nd October. The earliest sighting refers to the 14th June and the latest was on 11th November. No spring records were received.

Mediterranean Shearwater *Puffinus yelkouan*
Rare autumn passage migrant. Birds of the west Mediterranean form *P. y. mauretanicus* formerly considered a race of Manx Shearwater *P. puffinus* were recorded as follows:

> All records refer to the Bridges of Ross unless stated
> 1983: One on 15th October
> 1984: 10 between 9th and 23rd September, with a maximum of three on 9th September
> 1985: One at Castle Point, Kilkee on 1st September
> 1986: Singles on 18th and 22nd October
> 1987: 13 between 30th July and 6th October, with a maximum of four on 2nd September
> 1988: One on 23rd September
> 1990: One off Loop Head on 11th August, 21 from 12th August to 19th September, with a maximum of four on 12th August
> 1991: Singles on 7th and 21st September and two on 16th September

Manx Shearwater *Puffinus puffinus*
Regular on passage at the Bridges of Ross from March to October. The annual peak hourly movements are summarised as follows;

> 1982: 357 on 25th May
> 1983: 972 on 2nd September
> 1984: 3000 on 4 th September
> 1985: 5652 on 5th August
> 1986: 339 on 2nd September
> 1987: 2280 on 25th August
> 1988: 3120 on 23 rd August
> 1989: 1500 on 12th August
> 1990: 2000 on 15th August
> 1991: 660 on 24 th July

Early spring counts worthy of note from elsewhere in Clare include 20 at Black Head on 25th March 1989 and 20 at Kilkee on 19th March 1991.

Little Shearwater* *Puffinus assimilis*
Extremely rare autumn visitor.

1985: One at Bridges of Ross on 31th August (C.C.Moore)
1991: One at Bridges of Ross on 30th September (T.Tarpey)

These are the first and second records for Clare of this rare seabird.

Wilson's Petrel* *Oceanites oceanicus*
Rare autumn visitor.

1985: One at Bridges of Ross on 18th August (P.Buckley)
1990: One at Bridges of Ross on 15th August (K.Mullarney et al)

These are the third and fifth Irish land records of this mainly pelagic species, and the first Clare records (see description).

Storm Petrel *Hydrobates pelagicus*
Regular passage migrant and summer visitor, possibly breeding on Mattle Island. Though there are some instances of heavy autumn passage at the Bridges of Ross, the typical pattern is erratic with Storm Petrel sometimes being outnumbered by Leach's Petrel. The strongest passage is in early autumn with a peak of 500 per hour on 15th August 1990. Birds have been recorded in the inner Shannon Estuary on a number of occasions following storms, with an exceptional 'wreck' occurring on 3rd September 1983 when 16 birds were seen at Shannon Airport Lagoon.

Leach's Petrel* *Oceanodroma leucorhoa*
Regularly seen during autumn seawatching at the Bridges of Ross. The timing of this passage is summarised below:

	August		September		October		TOTAL
	1st - 15th	16th - 31st	1st - 15th	16th - 30th	1st - 15th	16th - 31st	
1982	-	-	-	4	1	2	7
1983	-	-	24	38	18	2	82
1984	-	-	240	568	2	60	870
1985	1	5	70	-	23	-	99
1986	-	-	1	-	1	53	55
1987	-	-	141	-	-	-	141
1988	-	-	570	92	84	-	746
1989	-	14	-	38	5	-	57
1990	-	2	364	51	30	-	447
1991	-	1	40	102	15	15	173**
Total	1	22	1450	893	179	132	2677

With 87% of birds recorded in September, the prime time for Leach's is heavily weighted towards the middle three weeks of this month. The heaviest daily passage occurred on the evening of 11th September 1988 when 486 were seen in just four and a half hours. Interestingly there is only one spring record from this site of eight in April 1985.

Other Clare records were all in 1988; one on 25th September and two on 1st October at Kilbaha and one long-dead corpse at Lough O'Donnell on 9th October. The only Limerick records were of two birds found dead at Auginish Island on 22nd September 1983 and one at Charleville Lagoons on 24th December 1989. This inland winter record following southerly gales was associated with a widespread wreck of this species in southern Britain.

Gannet *Sula bassana*
Regular all along the Clare coast even in winter. Peak passage off the Bridges of Ross include 1140 per hour on 17th October 1982 and 1400 per hour on 7th October 1987. Storm-blown birds have turned up at Shannon jetty and well inland at Meanus, Limerick.

Cormorant *Phalacrocorax carbo*
Resident. Common throughout the year on the coast, particularly in the Shannon Estuary and along the river Shannon, and also at certain inland lakes. Notable counts include a flock of 200 passing Mellon Point on 13th November 1982, 160 at St. Thomas's Island on 12th October 1982, 216 on a Shannon Estuary count on 30th December 1984, 43 at Lough Gur on 27th October 1985, 50 at Meelick Bay, Lough Derg on 20th August 1988 and 120 at Black Head on 1st November 1989.

According to the 1985-86 breeding survey (MacDonald 1987), there were 50 breeding pairs in Clare. This represents a drop of 50% on the Operation Seafarer 1969-70 survey, but a rise may have taken place since, as a trip to Mattle Island off Quilty on 19th May 1990 produced a count of 60 nests. No breeding was reported from Limerick.

Shag *Phalacrocorax aristotelis*
Common resident. Numerous breeding colonies on the Clare coast from which 57 nests were reported from the Cliffs of Moher in 1982 and 25 nests on Mattle Island, Quilty in 1990. Counts from outside the breeding season were 40 at Liscannor on 3rd January 1987 and 25 at Loop Head on 21st September 1991.

Bittern* *Botaurus stellaris*
Rare migrant.
> 1983: One was present at a site in Clare from 24th April to 22nd May
> (V.Sheridan, J.Tottenham, R.Tottenham et al).

Little Egret* *Egretta garzetta*
Rare spring migrant. Only two records were received throughout the period.
> 1984: One at Shannon Airport Lagoon from 30th June to 29th July
> (P.Brennan et al)
> 1987: One at Shannon Airport Lagoon on 8th May (P.Brennan)**

Grey Heron *Ardea cinerea*
Widespread and common resident. Counts of over 20 include 24 at Westfields during the winter of 1982 and 27 at Aughinish, Foynes on 19th July 1982.

Mute Swan *Cygnus olor*
Common resident. There was an exceptional moult concentration of 233 at Shannon Airport Lagoons on 4th August 1982. Other sizable counts include 162 at Lough Atedaun on 1st December 1985, 70 at Lough Gur on 26th December 1985, and 62 at Longpavement Marsh on 12th December 1982.

Bewick's Swan *Cygnus columbianus*
Winter visitor which appears to have suffered a significant population decline throughout the region compared to the 1970s. The only substantial flock recorded was 37 at Ballyallia Lake on 15th December 1984. Other counts were 16 at Shannon Airport Lagoons on 8th December 1987, 10 at Longpavement Marsh on 13th February 1982 and eight at Rathcannon, Bruff on 16th February 1986.

Whooper Swan *Cygnus cygnus*
Widespread winter visitor to many lakes, turloughs and the Shannon Estuary. As part of a national census during January 1986, 318 birds were located at 17 sites in Clare and 111 birds at five sites in Limerick. The largest concentration in Clare occurs along the Corofin lake chain with 184 at Lough Atedaun on 24th March 1985 and 168 at Lough Cullaun on 13th January 1991 being peak site counts for this area. Lesser numbers frequent the Burren turloughs and east Clare lakes. The most significant Shannon Estuary count was 70 at Islandavanna on 4th March 1989. Peak site counts for Limerick were 62 at Lough Gur on 22nd March 1987 and 49 at Longpavement Marsh on 12th December 1982. Interestingly a bird summered at Shannon Airport Lagoon in 1982.

Bean Goose* *Anser fabalis*
Very rare winter visitor, with only one record received during the period.
> 1982: Seven at Clarina, Limerick from 23rd to 31st January (R.Kiely)

This is the first definite record for the region of this species since 1903.

Greenland White-fronted Goose *Anser albifrons*

Regular winter visitor. A number of distinct flocks winter in the region. A flock frequents both sides of the Shannon Estuary and also the Fergus Estuary. Counts received include 20 in the Fergus Estuary on 13th January 1990 and 20 at Mellon Point on 19th December 1982. A south Clare flock is normally based at Tullagher Lake where 41 were present on 16th February 1983. Another flock moves between Liscannor Bay/Inagh Estuary and the east Burren chain of lakes. The peak count of this flock was 60 at Lough Bunny on 14th January 1990. An east Clare flock moves between Loughs Graney and O'Grady and Lough Derg, the largest count being 28 at Lough Graney on 24th January 1989.

The Winter Wetlands Survey 1984/85 - 1986/87 produced the following flock data:

Shannon/Fergus Estuaries	53
Tullagher	41
Liscannor/south Burren	54
East Clare	65
Total	213

Greylag Goose *Anser anser*

Regular winter visitor, now mainly confined to the Maigue Estuary/Mellon Point area. Counts at Mellon Point include 110 on 27th January 1985 and the same number also counted there in March 1986 during the Greylag Goose Census (Merne 1986). Surprisingly, during the 1986 Greylag Goose Census three birds were also noted at Ballyallia Lake. Sightings on the Clare side, probably part of the Maigue flock, include 21 at Hurlers Cross on 30th November 1986, 13 at Shannon Town on 8th April 1991 and six at Clenagh on 6th January 1990.

Barnacle Goose *Branta leucopsis*

Regular winter visitor. Flocks occur along the west Clare coast, mainly at Mutton Island where 343 were counted on 1st January 1982 and in more variable numbers at Illaunaneraun where 200 were present on 6th April 1988. Also regularly recorded in smaller numbers at Loop Head on autumn passage from as early as 5th October 1984.

Brent Goose *Branta bernicla*

Regular winter visitor to the lower Shannon Estuary (mainly to Poulnasherry and Bunclugga Bay) and also to the Ballyvaughan Bay /Lough Muree area. A total of 160 was present at Ballyvaughan on 28th March 1984. Also regularly recorded at the Bridges of Ross on autumn passage as early as 4th September 1986.

Brent Geese *(Ken Preston)*

Shelduck *Tadorna tadorna*
Summer resident and winter visitor. Breeds in considerable numbers throughout the Shannon Estuary and also on the Clare coast, as illustrated by a count of 21 juveniles at Clonderlaw Bay on 27th August 1983. Birds begin to return from moult migration in bulk during December. The Winter Wetlands Survey 1984/85- 1986/87 produced a Shannon Estuary count of 502 birds. At Shannon Airport Lagoon 175 were present on 23rd February 1986. Notable counts from outside the Shannon Estuary include 70 at Bell Harbour on 13th October 1989 and nine inland at Muckanagh Lough on 29th November 1986.

An interesting cameo is the annual presence of up to four pairs far inland at Charleville Lagoons in spring, with birds remaining to breed successfully only in 1988 when one pair fledged seven young.

Wigeon *Anas penelope*
Widespread winter visitor with birds beginning to arrive by late August. The Winter Wetlands Survey showed a marked decline in numbers in the Shannon and Fergus Estuaries. The count of 2229 represents a 71% decline on the 1972/73 - 1976/77 figures of 7816. Other notable winter maxima include 1390 at Ballyallia Lake on 24st January 1987, 1100 at Lough Inchiquin on 28th November 1990 and 1582 at Lough Gur on 22nd February 1987. Summer records include females at Aughinish, Limerick on 19th May 1982 and at Lough Gur on 6th June 1987, and males at Shannon Airport and Charleville Lagoons on 19th June 1983 and 22nd June 1986 respectively, and a pair at Shannon Airport Lagoon on 18th May 1987.

13

American Wigeon* *Anas americana*
Rare nearctic vagrant.
1991: An adult drake at Lough O'Grady from 19th November 1991 into 1992 (A.D. Smith).

This is only the second record of this species for both Clare and the region.

Gadwall *Anas strepera*
Winter visitor.   Numbers seem to have undergone a marked decline at Ballyallia Lake from a 1980 maximum of 200 down to a 1982-91 peak of 60 on 11th November 1990.   Other significant site maxima were 42 at Shannon Airport Lagoon on 12th January 1983, 38 at Lough Gash on   26th December 1990 and 32 at   Lough Gur on 27th January 1990.   The only breeding record comes from Lough Gur in 1987, a first for the county, although a pair was present at Shannon Airport Lagoon throughout May 1987.

Teal *Anas crecca*
Common winter visitor and   passage migrant.   Breeds in small numbers in both counties.   A Winter Wetlands Count of the Shannon Estuary on 30th December 1984 produced a total of 2303, which represents a 53% decrease compared to the   1972/73 - 76/77 figure of 2735.   Some peak counts from outside the Shannon Estuary include 2252   at Ballyallia Lake on   24th January 1987, 1000 at Lough Inchiquin on 24th November 1990 and 737 at Charleville Lagoons   on 14th December 1986.   Only summer records are of a female at Lough Gur on 6th June 1987 and a male at Charleville Lagoons on 22nd June 1986.

Birds of the American race *A. c. carolinensis** were recorded as follows:
1982:   A male at Kilrush on 18th February (P.Buckley)
1983:   A male at Aughinish Island, Limerick on 29th October (E.E.H.Jones)
1988-90:   A male at Westfields from 22nd November 1988 to 22nd January 1989, with probably the same bird present again on 21st January 1990 (T.Tarpey).

Mallard *Anas platyrhynchos*
Very widespread resident breeding in good numbers, and also a winter visitor.   A total of 456 was counted throughout the Shannon and Fergus Estuaries on 30th December 1984.   Some other peak counts include 542 at Lough Gur on   21st December 1986, 350 at Charleville Lagoons on   4th August 1986, 610 at Ballyallia Lake on 24th January 1987, 350 at   Shannon Airport Lagoon on 31st July 1990 and 200 off the coast at Corranroo Bay on 13th September 1990.

Pintail *Anas acuta*

Regular winter visitor in small numbers to Clare, though much scarcer in Limerick. Some peak counts include 56 at Tarbert Bay on 11th January 1982, 50 at Shannon Airport Lagoon on 7th February 1987, 32 at Clonderlaw Bay on 27th February 1982 and 23 at Ballyallia Lake on 24th January 1987. Birds were recorded present in two squares in Clare during the *New Breeding Atlas* 1988-91.

Garganey* *Anas querquedula*

Rare continental European migrant (only one record received).

1988: A female/immature at Lough O'Donnell on 24th August (R.A.Frost)

This is only the second Clare record of this species.

Shoveler *Anas clypeata*

Winter visitor and passage migrant. Ballyallia Lake is a site of international importance for this species with 793 present there on 24th January 1987. Other site maxima of note were 170 at Muckanagh Lough on 31st January 1982, 200 throughout the Shannon Estuary on 30th December 1984, 203 at Lough Gur on 21st December 1986 and 198 at Charleville Lagoons on 12th October 1986. Summer records include three drakes at Westfields in July 1983 and four females and a male at Charleville Lagoons on 22nd June 1986.

Pochard *Anthya ferina*

Winter visitor and possibly a scarce resident, as birds were recorded in three squares during the 1988-91 Breeding Atlas. Some peak counts include 239 at Lough Gur on 21st December 1986, 160 on River Shannon, Limerick City on 13th January 1982, 156 at Charleville Lagoons on 23rd November 1986 and 132 at Fenloe Lake on 11th January 1986. Summer records include four females and one male at Lough Gur on 6th June 1987.

Ring-necked Duck* *Aythya collaris*

Rare nearctic vagrant.

1982: A male at Lough Gur on 1st January (C.W.Hutchinson)
A male at Lough Gur on 28th March (C.C.Moore)
1987: A male at Charleville Lagoons on 20th June (T.Mee)

These are the second and third records for Limerick of this rarity. The 1982 records probably refer to the same bird.

Ferruginous Duck* *Aythya nyroca*
Rare continental European winter visitor.
1991: An adult male at Lough Gur on 26th January (T.Mee)

This is the first record for the region of this species.

Ferruginous Duck *(John Murphy)*

Tufted Duck *Aythya fuligula*
Resident and winter visitor, breeding mainly in the eastern part of the region. Some peak counts include 488 at Lough Derg (Clare section) on 28th February 1987, 429 at Lough Gur on 27th October 1985, 266 at Ballyallia Lake on 24th January 1987 and 220 at Bunlicky Lake, Mungret in January 1986.

Scaup *Aythya marila*
Winter visitor and passage migrant mainly to the Shannon Estuary. A total of 273 was recorded on 30th December 1984 during a Shannon Estuary count. Other significant Shannon Estuary counts include 138 at Tarbert Bay on 11th January 1982, 155 at Clonderlaw Bay on 29th October 1984 and 200 at Islandavanna on 10th May 1990. Two females remaining at Clonderlaw Bay until the 15th June 1983 were the only summer records. One or two birds reported regularly in winter from Westfields, Charleville Lagoons, Ballyallia Lake and Lough Donnell.

Common Eider *Somateria mollissima*

Rare winter visitor and passage migrant. The following records have been received:

    1983: A female at Black Head from 13th to 20th February
    1988: A pair in Clare in September
          A female at Ross Bay, Loop Head on 12th November
    1990: A female at Kilbaha from 13th to 15th November
    1991: A female shot in the Fergus Estuary in January

These are the first records for the region of this sea-duck, and probably originate from the expanding breeding population along the north-west coast of Ireland. This species was also recorded as seen during the *New Breeding Atlas 1988-91* in the Kilrush area.

Long-tailed Duck *Clangula hyemalis*

Regular winter visitor to the Clare coast and occasionally inland. A sample of counts include 12 at Flaggy Shore on 31st December 1988 and 12 at Liscannor Bay on 13th January 1983. Inland records include a pair at Lough Atedaun on 24th January 1987 and at Charleville Lagoon, a female on 12th November 1988 and an immature male from 20th October to 25th November 1989.

Common Scoter *Melanitta nigra*

Passage migrant and winter visitor off the Clare coast. Most regular concentrations at Liscannor Bay where 75 were present on 18th April 1982. Other counts include 35 at The Rine, Ballyvaughan on 12th February 1989 and 30 at the Flaggy Shore on 31st December 1988. The only Shannon Estuary record was of a male at Clonderlaw Bay on 1st October 1983. Regularly recorded on autumn passage at the Bridges of Ross where 23 passed on 15th September 1991. The only summer records involved a pair east of Mountshannon, Lough Derg in June 1983, and an adult male seen off Illaunmore, Lough Derg on 27th June 1987.

Surf Scoter* *Melanitta perspicillata*

Nearctic vagrant.

    1984: Three adult males and a first-winter male at Lahinch from 28th March to
          7th April (M.Donahue, K.Mullarney, T.Tarpey, et al).
          Adult male at Ballyvaughan Bay on 8th April (K.Mullarney)
    1985: Adult male at Lahinch on 23rd March (T.Tarpey)

Velvet Scoter *Melanitta fusca*

Rare winter visitor.

    1983: One at Aughinish, Limerick on 19th October
    1988: One at Ballyvaughan Bay on 17th January
    1991: Two at Finvarra on 28th January

These are the first records for the region of this rare sea-duck.

Goldeneye *Bucephala clangula*
Winter visitor to the Shannon valley above Limerick City including Lough Derg, and also the east Burren lakes. Peak counts for the different sites are 147 at Lough Derg (Clare side) on 28th February 1987, 86 at Shannon Headrace on 28th February 1987, 80 at Fenloe Lake on 10th December 1990, 51 at Lough Bunny/George/Muckanagh on 31st January 1982 and 46 at Clonlea Lake on 11th January 1986. Very scarce in Limerick where the only count of note was eight at Charleville Lagoons on 25th November 1989. Interesting summer records include a male at Fenloe Lake on 10th May 1990 and a female at nearby Kilkishen Lake on 4th June 1990.

Red-breasted Merganser *Mergus serrator*
Passage migrant, winter visitor and scarce summer resident. Most plentiful in winter in the Ballyvaughan Bay area where a peak of 80 was recorded at Corranroo on 11th November 1989. Present in smaller numbers throughout the Shannon Estuary right up to Limerick City, where the only significant count was 10 at Shannon Airport Lagoon on 26th September 1987. Breeds on Lough Derg and occasionally on some of the smaller Clare lakes, but the only record received was of a family party of six at Meelick Bay, Lough Derg on 20th August 1988.

Goosander* *Mergus merganser*
Rare winter visitor.
> 1989: A male at Lough Cullaun on 29th October 1989 (R.Ackerly)**

This is only the second record for the region of this species.

Ruddy Duck *Oxyura jamaicensis*
Vagrant from now well established populations in Northern Ireland and Britain. An adult male that was present at Lough Gur on 2nd June 1987 moved to Charleville Lagoons and stayed from 20th June to 25th July. A male returned to summer at Charleville Lagoons for the following four years, 1988 to 1991 inclusive. These records almost certainly refer to the same bird and are the first records for the region of this species which originally hails from North America but has become naturalised into the wild in Britain from escaped stock.

Marsh Harrier* *Circus aeruginosus*
Three records were received for this species:
> 1981: A male was present at a locality in Clare from 16th to 25th May and
> constructed a nest platform (P.Roscoe)
> 1989: A juvenile at Shannon Airport Lagoon on 17th October (P.Brennan)**

These are only the second and third records for the region of this rare migrant.

Hen Harrier *Circus cyaneus*

Scarce resident and passage migrant. Breeding is confined to a number of upland areas in both counties, where this species has become increasingly dependent on the availability of young forestry plantations for suitable nesting sites. In autumn and winter it is much more evident at lowland wetlands such as the east Burren wetlands and along the coast. Regularly recorded at Loop Head during autumn coverage including one interesting report of a female being violently mobbed for over five minutes by a Grey Heron on 25th August 1987.

Sparrowhawk *Accipter nisus*

Widespread and relatively common resident.

Buzzard* *Buteo buteo*

Rare vagrant.
> 1988: One at Loop Head on 31st October (P.Brennan, G.Donnelly)
> 1991: One at Kilkee on 3rd May (C. Meehan)**

These are the first and second records of this species for the region.

Osprey* *Pandion haliaetus*

Rare migrant.
> 1984: One at Lough Derg from 4th to 23rd June (C.Moriarty, J.O'Flaherty)

This is the second Clare record of this species and the first for this century.

Kestrel *Falco tinnunculus*

Widespread resident. Regularly recorded at Loop Head during autumn coverage with four being the maximum daily count.

Merlin *Falco columbarius*

Scarce resident, passage migrant and winter visitor. Bred at two known sites in Limerick during the period, and bred in a number of favoured areas in Clare. More numerous in winter when birds are regularly seen at sites along the Shannon Estuary such as Shannon Airport Lagoon. Regular on autumn passage at Loop Head where there are almost daily records during October and peak counts of at least three birds in a day.

Hobby* *Falco subbuteo*

Rare passage migrant.
> 1987: One at Loop Head from 1st to 7th November (T.Mee)
> 1989: One at Lough Gur on 18th August (T.Mee, T.Tarpey)

These are the first and second records of this species for the region.

Peregrine *Falco peregrinus*
Scarce resident mainly confined in breeding terms to the Clare sea cliffs but does breed inland in the region, including Limerick. Regularly recorded in winter at wetland sites in both counties as well as in more unexpected locations, such as one bird which resided on the steeple of St. John's Cathedral, Limerick during the 1983/84 winter. Regularly recorded during autumn coverage at Loop Head with a maximum count of three birds.

Red Grouse *Lagopus lagopus*
Scarce resident. Breeds in small numbers in a number of relict heather moors in both counties, but has suffered greatly from afforestation and land reclaimation. Notable counts include 14 at Long Mountain, Ballyhouras on 13th October 1985, seven at Woodcock Hill on 10th November 1986, and four at Slieve Barnagh on 2nd October 1982.

Grey Partridge *Perdix perdix*
Very rare resident. The *New Breeding Atlas 1988-91* shows a change in distribution within the region compared to the *1968-72 Breeding Atlas*, summarised below:

	Squares present	
	1988-91	1968-72
Clare	-	2
Limerick	2	-

Quail *Coturnix coturnix*
This species was recorded during the *New Breeding Atlas 1988-91* at an east Clare location.

Pheasant *Phasianus colchicus*
Very widespread resident, even in the more barren areas of west Clare.

Water Rail *Rallus aquaticus*
Widespread resident and winter visitor to various lowland marshes. During the summer of 1985, 26 birds were trapped and ringed at Shannon Airport Lagoon but this species is undoubtedly underrecorded due to its secretive behaviour. Good numbers have been reported from Westfields and Lough Gur.

Spotted Crake* *Porzana porzana*

Rare passage migrant.

1989: A juvenile, found dead at Loop Head on 4th October (T.Tarpey et al).

This is the third Clare record of this species.

Corncrake *Crex crex*

1982 marked the final breeding season for this endangered species in the Shannon Town area, one of the last strongholds for this species in Clare. During the 1988 IWC survey there were three records from Limerick and 12 from Clare. The majority of records were single calling birds on passage with more than one bird recorded together at only a single western site in each of the two counties. The *New Breeding Atlas 1988-91* produced a similarly dismal picture, with two confirmed breeding records for Clare and none for Limerick.

Moorhen *Gallinula chloropus*

Widespread and common resident, with notable counts of 30 at Westfields on 6th February 1982, and 17 at Ballyallia Lake on 23rd November 1985.

Coot *Fulica atra*

Widespread resident and winter visitor. Some peak counts include 928 at Lough Gur on 18th January 1987, 570 at Lough Atedaun on 1st December 1985, 488 at Ballyallia Lake on 23rd November 1985 and 397 at Lough Derg (Clare side) on 28th February 1987.

Common Crane* *Grus grus*

Rare vagrant.

1986: An adult at New Quay from 5th March to 3rd April (N.Walsh)

This is the first record for the region of this species. This record comes courtesy of Bus Éireann as the bird was first spotted from the Ballyvaughan bus!

Oystercatcher *Haematopus ostralegus*

Widespread winter visitor and less numerous summer resident. Breeds in a few localities along the Clare coast, and probably on some of the islands in the Shannon Estuary. In winter there is a large population scattered along the Clare coast and throughout the Shannon Estuary. A total of 1541 birds was recorded along the west Clare coast in December 1987, representing 35% of the non-estuarine Irish west coast population (Green et al 1988). Some notable counts include 250 at Loop Head on 10th November 1985, 200 at Hags Head on 19th March 1989 and 199 on a Shannon Estuary count on 30th December 1984. One inland record of a single bird at Charleville Lagoon on 11th August 1985.

Black-winged Stilt* *Himantopus himantopus*
Rare vagrant.

1987: A male at Ennistymon from 22nd to 26th April (G.Pearson)

This is the first Clare record and the second for the region.

Ringed Plover *Charadrius hiaticula*
Summer resident, passage migrant and winter visitor. Breeds on some of the Burren lakes and along the Clare coast and Shannon Estuary. A total of 607 was recorded along the west Clare coast in December 1987 (Green et al 1988). Some peak counts include 500 at Aughinish, Clare on 25th January 1991, 210 at Seafield on 2nd January 1982, and 91 at Tarbert Bay on 11th January 1982.

Dotterel* *Charadrius morinellus*
Rare migrant.

1988: A juvenile at Loop Head on 9th October (T.Mee et al)

This is the first record for the region of this species.

Golden Plover *Pluvialis apricaria*
Widespread and numerous winter visitor and passage migrant. Large roosting flocks occur on exposed mudflats in the Shannon Estuary. A total of 8090 was recorded in a Shannon Estuary count on 30th December 1984 (including counts of 2500 at Greenish Island and 2500 at Clenagh which were significantly greater than previous censuses). Other peak counts include 2000 at Robertstown on 1st October 1983, 3000 at Shannon Airport Lagoon in late December 1988 and 2000 there on 27th March 1989. Inland counts include 1000 at Lough Atedaun on 23rd January 1989 and 930 at Rathcannon, Bruff on 30th November 1986.

Grey Plover *Pluvialis squatarola*
Winter visitor and passage migrant. Occurs throughout the Shannon Estuary and along the Clare coast in small flocks. A total of 90 was recorded in a Shannon Estuary count on 30th December 1984 and 148 were recorded along the west Clare coast in December 1987. Other peak counts include 100 at Shannon Airport Lagoon on 17th March 1989 and 50 at Islandavanna on 29th January 1989. Four occurred inland at Charleville Lagoons on 1st December 1985 following severe gales.

Lapwing *Vanellus vanellus*
Winter visitor, passage migrant and summer resident in Clare, and to a lesser extent, in Limerick. The maximum numbers encountered in the Shannon Estuary during the Wetland Survey 1984/85-86/87 were 13029, which represents a significant increase on previously recorded levels for the estuary. Significant counts from other areas include 2000 at Charleville Lagoons on 12th January 1990 and 1100 at Loop Head on 14th November 1987.

Knot *Calidris canutus*
Winter visitor and passage migrant. Significant counts include 1500 at Shannon on 19th January 1991, 756 in the Shannon Estuary on 30th December 1984 and an early peak of 300 at Iniscullen, Shannon on 4th September 1983. Regularly recorded on autumn seawatches at the Bridges of Ross with a peak of 40 on 16th September 1991.

Sanderling *Calidris alba*
Passage migrant and winter visitor to the Clare coast and the lower reaches of the Shannon Estuary. A total of 161 was recorded along the west Clare coast in December 1987 (Green et al 1988). Peak counts include 100 at Seafield on 27th January 1990, 40 at Kilkee on 27th January 1990 and five at Shannon Airport Lagoon on 4th January 1986.

Little Stint *Calidris minuta*
Scarce but regular passage migrant. Regularly recorded in autumn at Shannon Airport Lagoon with autumn peaks summarised as follows:

1982	1983	1984	1985	1986	1987	1988	1989	1990	1991
4	4	-	11	1	-	2	-	3	-

Records from other sites include three at Lough O'Donnell on 5th October 1983, seven at Seafield on 5th October 1983, and two at Bridges of Ross on 4th September 1988. The only Limerick records comes from inland at Charleville Lagoons where single birds were present on 28th August 1987 and 17th September 1991. The only spring record is of a single bird at Seafield on 3rd March 1990.

White-rumped Sandpiper* *Calidris fuscicollis*
Rare nearctic vagrant.
1990: One at Shannon Airport Lagoon on 15th September 1990 (P. Brennan).**

This is the second record for this site and for the region, of this species.

Pectoral Sandpiper *Calidris melanotos*
Rare nearctic vagrant.
1983: One at Aughinish, Limerick on 18th August.
1987: A juvenile at Lough O'Donnell on 13th September.

This is the first record for Limerick and the third record for Clare of this species.

Curlew Sandpiper *Calidris ferruginea*
Scarce but regular passage migrant. Regular at Shannon Airport Lagoon in autumn with peak counts summarised as follows:

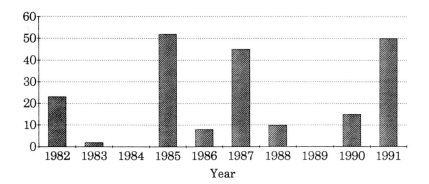

Records from less well-watched sites in Clare include two at Kilkee on   5th October 1983, 17 at Seafield on 17th October 1983, one at Lough O'Donnell on 3rd September 1988, four at Bridges of Ross on 23rd September 1988 and five at Moyasta on 29th September 1990.   Records from Limerick include two at Robertstown on 1st October 1983, one at Loughill on 2nd October 1983 and nine inland at Charleville Lagoons on 24th September 1991.   The earliest record is of two birds in summer plumage at Shannon Airport Lagoon on 28th July 1991.   An exceptionally late record was of two birds trapped and ringed at Shannon Airport Lagoon on 27th November 1988.

Purple Sandpiper *Calidrus maritima*
Winter visitor to the Clare coast with Quilty Bay, where rotting kelp beds on a sandy beach attracts a large flock, being the most important site for the species on Ireland's west coast. A total of 242 was recorded along the Clare coast in December 1987 which represents 44% of the non-estuarine Irish west coast population (Green et al 1988). Peak counts include 202 at Seafield on 14th January 1983, and 230 there on 3rd March 1990. Counts from elsewhere on the coast include 30 at Bridges of Ross on 10th November 1985, and 10 at Hags Head on 19th March 1989.

Dunlin *Calidris alpina*
Scarce resident, widespread and numerous winter visitor and passage migrant. May breed around Lough Derg and also at some of the Burren lakes. A total of 13329 was recorded during a Shannon Estuary count on 30th December 1984, including 3850 at Clenagh. This represents a decline of more than 50% on the Shannon Estuary compared to 1973/74 counts. Some peak site counts include 6000 at Shannon Airport Lagoon on 14th January 1991, 1941 at Poulnasherry Bay on 15th February 1982 and 1520 at Tarbert Bay on 11th January 1982. A total of 1523 was recorded along the west Clare coast in December 1987.

The highest spring passage count at Shannon Airport Lagoon was 1120 on 30th April 1983. Of 81 birds trapped and ringed there in May 1987, 12 were of the Greenland race *C. a. arctica* and the balance were of the race *C. a. schinzii* (probably from the Iceland or south-Greenland populations). Included in this catch was a bird originally ringed in Mauretania, West Africa and another which was found later on the nest in Iceland.

Broad-billed Sandpiper* *Limicola falcinellus*
Rare passage migrant.
    1978: One at Charleville Lagoons on 2nd June (the late S.C. Johnson)

This record (previously rejected) is the fourth Irish record and the first for the region of this species.

Buff-breasted Sandpiper* *Tryngites subruficollis*
Rare nearctic vagrant.
    1990: Two at Bridges of Ross on 6th September (K.Mullarney, et al)
        One at Bridges of Ross on 26th October (T.Mee)

These are the second and third records for the region of this species.

Ruff *Philomachus pugnax*

Scarce passage migrant. Regular on autumn passage at Charleville Lagoons where birds have been recorded from 26th July to 22nd November with peak counts of eight on 10th August 1986 and an unprecedented influx of 22 on 30th August 1991. Other counts include five at Aughinish, Limerick on 20th August 1982, two at Shannon Airport Lagoon on 17th August 1986 and five there again on 24th August 1987, four at Bridges of Ross on 24th September 1988, five at Loop Head on 26th August 1989 and four at Flaggy Shore on 1st November 1989. The only spring sightings were of two at Shannon Airport Lagoon on 29th April 1987 and three at Charleville Lagoons on 8th April 1990.

Jack Snipe *Lymnocryptes minimus*

Under-recorded winter visitor. All the records refer to single birds:

    1983: Labasheeda on 10th January
           Shannon Airport Lagoon on 15th January.
    1984: Foynes on 30th December
    1988: Kilbaha on 29th October
    1991: Fenloe Lake on 19th January

Snipe *Gallinago gallinago*

Widespread and numerous winter visitor and summer resident. On 26th March 1983, 279 were counted leaving Westfields at dusk to feed at night further down the Shannon Estuary. Other notable counts include 201 at Clonlea Lough on 11th January 1986, 129 at Charleville Lagoons on 8th February 1987 and 112 at Shannon Airport Lagoon on 15th January 1983. The extent of migration at Loop Head is hard to quantify but 54 were present on 26th August 1988.

Dowitcher sp.* *Limnodromus scolopaceus/griseus*

Rare nearctic waders.

    1983: A dowitcher, not specifically identified, at Lough O'Donnell on 5th October
           (P.Buckley)

This is the first occurence for the region of a member of this family of waders.

Woodcock *Scolopax rusticola*

Widespread winter visitor and localised summer resident. Only notable counts were five at Ballycasey on 1st January 1990, 10 at Fenloe Lake on 17th February 1990 and 10 at Cratloe on 2nd July 1991. Recorded on autumn passage at Loop Head from late October onwards, with a peak of five on 6th November 1988. Three were attracted to Loop Head Lighthouse on the night of 30th October 1989.

Black-tailed Godwit *Limosa limosa*

Winter visitor, passage migrant and non-breeding summer resident. The Shannon and Fergus Estuaries are of international importance for this species. The total peak count for both estuaries, recorded during the spring of 1987 as part of the Wetlands Survey 1984/85-86/87, was 5370. Highest concentrations occur during April and early May at their apparently traditional pre-migration collection area in the Fergus Estuary, although an unusually early spring concentration of 1320 birds was noted at Shannon Airport Lagoon on 25th February 1987. More typical counts include a spring peak of 1500 at Islandavanna on 27th April 1989 and an autumn peak of 1200 at Shannon Airport Lagoon on 27th September 1988. Regular inland at Ballyallia Lake where 200 were present on 8th December 1990.

Bar-tailed Godwit *Limosa lapponica*

Winter visitor, passage migrant and non-breeding summer resident. A total of 651 was recorded during a Shannon Estuary count on 30th December 1984, including 215 at Shannon, 220 at Clenagh and 150 at Ing. 650 were present at Shannon Airport Lagoon on 23rd December 1985. Other notable counts from the west coast include 14 at Lough O'Donnell on 3rd September 1988 and 30 at Bridges of Ross from 19th to 23rd September 1988.

Whimbrel *Numenius phaeopus*

Numerous passage migrant. A substantial spring movement takes place throughout the region. The first birds tend to reach the mid-west around 20th April and passage tends to fizzle out around mid May. An interesting feature of this spring movement takes place at Charleville Lagoons, which acts as a traditional stop-over site, with a high turnover of birds using certain empty lagoon beds as a night roost throughout the migration period.

Some notable spring counts were 419 at Shannon Airport Lagooon on 28th April 1987, 200 at Westfields on 27th April 1983, 190 at Clonderlaw Bay on 21st April 1984, 180 at Corbally on 30th April 1989 and 248 at Charleville Lagoons on 4th May 1990. Flocks of over 100 are regularly seen flying west over Limerick City during this period, particularly during Munster Senior Cup finals at Thomond Park!

Autumn return passage tends to be a great deal less obvious with birds recorded inland in small numbers. The exception being along the west Clare coast where at the Bridges of Ross 302 were recorded on 28th July 1989 and 206 on 12th August 1989. The only winter record received is of one at Shannon Town on 5th January 1984.

Curlew *Numenius arquata*
Numerous and widespread winter visitor, passage migrant and localised summer resident. A Shannon Estuary count on 30th Decmber 1984 produced a total of 2039 birds which represents a 60% decline on 1974. A total of 402 was recorded along the west Clare coast in December 1987 (Green et al 1988). Other flock counts include 270 at Loop Head on 6th January 1983 and 390 at Charleville Lagoons on 15th March 1987.

Spotted Redshank *Tringa erithropus*
Regular passage migrant and scarce winter visitor. Most often seen in small flocks along the Shannon Estuary in spring and autumn, and occasionally one or two in winter. Notable counts were 10 at Aughinish, Limerick on 3rd October 1982, five at Clonderlaw Bay on 1st October 1983, 17 at Rogerstown on 1st October 1983, five at Corranroo, north Clare on 25th October 1989 and four at Manusmore, River Fergus on 27th April 1990. Winter records include two at Foynes on 21st January 1982, one at Poulaweala Creek on 30th December 1984 and singles inland at Charleville Lagoons on 8th February 1987, 12th November 1988 and 1st December 1991.

Redshank *Tringa totanus*
Widespread and numerous winter visitor, passage migrant and localised summer resident. The Shannon and Fergus Estuaries are still of international importance for this species where a total of 2377 was counted during a Shannon Estuary Count on 30th December 1984, including 490 at Clenagh and 406 at Islandavanna. Other notable counts include a pre-migratory concentration of 1000 at Shannon on 3rd April 1989 and counts of 445 at Tarbert Bay on 7th October 1982, 680 at Shannon on 29th October 1986 and 750 at Shannon on 24th July 1990. A total of 383 was recorded along the west Clare coast in December 1987 (Green et al 1988). Winters regularly inland at Charleville Lagoons where seven were present on 11th January 1987. Breeds in reasonable numbers on many of the Clare lakes particularly in north and east Clare with reports received of four pairs breeding at Knockaunroe, near Mullaghmore on 10th May 1988 and three pairs at Rinbarra Point, Lough Derg in 1987.

Greenshank *Tringa nebularia*
Winter visitor and passage migrant.   Occurs in small numbers throughout the Shannon Estuary and along the Clare coast.   Only notable counts were 11 at Tarbert Bay on 20th October 1982, 12 at Shannon on 13th November 1983, and eight at Moyasta on 29th September 1990.   Occurs occasionally inland with reports of five at Lough Gur on 22nd August 1984, three at Carron on 24th January 1987, three at Charleville Lagoons on 14th October 1990 and four at Lough Inchiquin on 14th December 1991.

Greater Yellowlegs* *Tringa melanoleuca*
Extremely rare nearctic vagrant.
1990: One at Moyasta on 29th September (J.Murphy)**

This, as yet unsubmitted record, is the first record for the region of this species.

Green Sandpiper *Tringa ochropus*
Passage migrant and winter visitor,   occurring mainly inland.   Charleville Lagoons is one of the premier sites in Ireland for this species.   Birds have been recorded here in every month with a latest spring record of two on 5th May 1991 and   an earliest return record of one on 22nd June 1986.   A notable autumn passage occurs during July and August, with a percentage remaining to over-winter.   Wintering numbers are determined by ground temperature/frost   conditions.

Autumn peak: 14 on 15th October 1988
Winter peaks: 16 on 24th December 1989

Individual bird-year totals are summarised as follows:

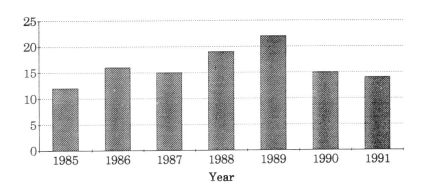

29

Elsewhere in the region, this species is surprisingly scarce with singles at Dernish Island on 27th December 1982, Askeaton on 30th March 1984, Shannon on three occasions and a pair at Loop Head on 27th August 1991.

Wood Sandpiper *Tringa glareola*
Passage migrant.   All records emanate from Limerick.
1982: One at Aughinish from 14th to 20th August and two on 21st August
1990: One at Charleville Lagoons on 29th August
1991: Singles at Charleville Lagoons on 12th May and 30th August

Common Sandpiper *Actitis hypoleucos*
Passage migrant and localised summer resident.   Breeding reported from some of the Burren lakes and Lough Derg, and the *New Breeding Atlas 1988-91* also confirmed breeding for the Foynes area.   Spring passage is light with single birds seen regularly in the Limerick City environs at Corbally and Westfields.   Also evident at Charleville Lagoons with a peak of four on 20th April 1989.   In autumn, records throughout the Shannon Estuary rarely exceeded single birds, with a peak count of four at Shannon Airport Lagoon on 30th July 1991.   Meanwhile, inland at Charleville Lagoons, birds were recorded as early as 22nd June and as late as 3rd October with main passage occuring during July and August.   Autumn maxima have varied here consistently between five and eight birds during the years 1988 to 1991.   Single birds were recorded in winter as follows: Aughinish on 30th December 1984; Bunlickey Lake, Mungret   in 1985/86; Ballyallia Lake from 1st January to 16th April 1989; and Limerick City   on 31st December 1991.

Turnstone *Arenaria interpres*
Winter visitor, passage migrant and non-breeding summer resident.   A total of 1069 was recorded along the Clare coast in December 1987 (Green et al 1988) with Quilty Bay being the most important coastal site for this species.   Also recorded regularly throughout the Shannon Estuary up to Limerick City.   Some peak counts include 235   at Seafield on 27th January 1990, 200 at Loop Head on 3rd October 1991, 60 at Poulnasherry Bay on 18th February 1982, and 80 at Shannon Airport Lagoon on 24th April 1983.

Wilson's Phalarope* *Phalaropus tricolor*
1984: A juvenile at Shannon Airport Lagoon, from 11th to 20th August
   (P. Brennan) **

This is the second Clare and site record of this rare nearctic vagrant.

Grey Phalarope *Phalaropus fulicarius*

Passage migrant with a regular autumn passage in variable numbers off the west Clare coast. Autumn passage at the Bridges of Ross is summarised as follows:

1982	1983	1984	1985	1986	1987	1988	1989	1990	1991
12	86	1093	28	25	47	154	3	122	250

Birds have been recorded here from 2nd August to 22nd October with 1984 being an exceptional autumn for this species. The only records from elsewhere in the region were one at Doolin on 3rd October 1982 and one inland at Charleville Lagoons following north-westerly gales.

Pomarine Skua *Stercorarius pomarinus*

Regular passage migrant off the Clare coast. The annual totals recorded at the Bridges of Ross were as follows:

1982	1983	1984	1985	1986	1987	1988	1989	1990	1991
25	273	41	21	30	30	233	2	76	137

The average weekly passage recorded at the Bridges of Ross during the period of this report is illustrated in the graph below. Typically the passage reaches a peak between mid September and mid October. The late October peak on the graph is primarily due to one exceptional movement of birds on the evening and the following morning of 15th/16th October 1983.

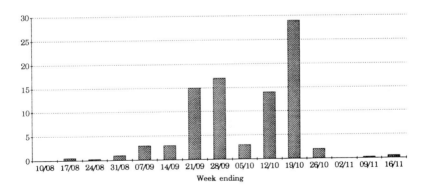

31

The highest recorded daily totals at the Bridges of Ross were:

- 111   on 15th October 1983
- 139   on 16th October 1983
- 59   on 6th October 1988
- 60   on 7th October 1988

There were no spring records for the period of this report. An age structure analysis of 181 birds observed at the Bridges of Ross between 23rd September and 7th October 1988 revealed the following composition:

	Juvenile	Sub-adult	Adult	Total
No. of birds	61	16	104	181
% of total	34	9	57	100

Arctic Skua *Stercorarius parasiticus*

Regular passage migrant off the Clare coast. The annual totals recorded at the Bridges of Ross were:

1982	1983	1984	1985	1986	1987	1988	1989	1990	1991
3	30	208	73	31	94	148	33	215	140

There was a sprinkling of records from other locations on the Clare coast and a single record from the Shannon Estuary, at Ringmoylan Pier on 4th September 1983. The autumn passage extends from early August into November, reaching a peak in early September. The highest recroded daily totals at the Bridges of Ross were:

- 42   on 15th August1985
- 98   on 5th September 1990
- 42   on 22nd September 1991

Arctic Skuas were recorded on 99 days during the autumn passage over the ten year period at the Bridges of Ross. This compares with 75 days for Pomerine Skua and 112 days for Great Skua. The average weekly passage for the period of the report is shown in the following graph.

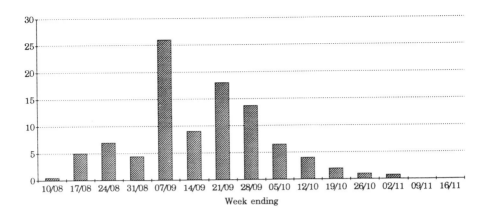

Week ending

Long-tailed Skua* *Stercorarius longicaudus*
Very scarce passage migrant off the Clare coast. A total of 58 birds was recorded, all at the Bridges of Ross, during the period of this report. The records are as follows:

1984: Five adults on 3rd September (T.Tarpey, E.Jones)
Singles on 24th September and 23rd November (P. Buckley)
1985: One adult on 14th September (T. Tarpey et al)
1988: Four juveniles on 2nd September (K.Mullarney)
One juvenile on 3rd September (P. Brennan et al)
Five adults and two juveniles on 28th September (V.Caschera et al)
Five adults on 7th October (A. Kelly, K. Mullarney)
1990: Four juveniles on 15th August (K. Mullarney)
One adult and five juveniles on 18th September (S. Farrell et al)
One juvenile on 6th October (P. Lonergan et al)
1991: One adult on 14th September (G.M.Cresswell, R.Martins & M.Evans)**
One adult and ten juveniles on 16th September (O. O'Sullivan et al)
One juvenile on 17th September (J.Holmes, M. Campbell)**
Two juveniles on 18th September (J.Holmes, M. Campbell)**
Three adults and four juveniles on 22nd September (P. Cummins, P. Brennan)**

The total weekly distribution of records over the period of the report is illustrated in the graph below.

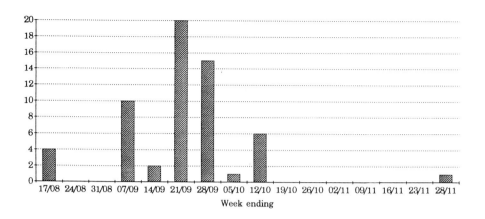

Week ending

These are the first series of records for Clare. This species has yet to be recorded in Limerick. The absence of records of juvenile birds in the earlier years is probably due to identification difficulties, rather than to any bias in the passage towards adult birds. The majority of records occurred between 1988 and 1991. Of these, 68% were identified as juveniles, reflecting increased observer awareness and interest in sea-watching in general.

Great Skua *Stercorarius skua*
Regular passage migrant off the Clare coast. The annual totals recorded during autumn passage at the Bridges of Ross are as follows:

1982	1983	1984	1985	1986	1987	1988	1989	1990	1991
63	132	436	177	42	265	335	68	287	382

The average weekly passage over the period of the report is shown on the graph below. The autumn passage extends from early August to mid November each year, usually reaching a peak in late September.

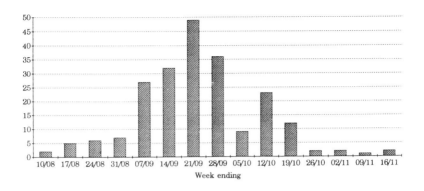

The highest recorded daily totals for the Bridges of Ross are 88 on 23rd September 1984, 86 on 5th September 1990 and 84 on 22nd September 1991. There are a number of winter records for Clare, most notably a count of 18 at the Bridges of Ross on 29th December 1987.

Mediterranean Gull *Larus melanocephalus*

Scarce autumn migrant. All records refer to the Bridges of Ross:
    1988: One at Bridges of Ross on 20th August
    1990: One at Bridges of Ross on 30th August
    1991: An adult at Bridges of Ross on 18th September

These are the first, second and third records of this species for the region.

Little Gull *Larus minutus*

Scarce passage migrant. Autumn passage at the Bridges of Ross is summarised as follow:

1982	1983	1984	1985	1986	1987	1988	1989	1990	1991
-	3	4	12	1	2	11	-	-	6

A nice scatter of records from elsewhere in the region are as follows:
    1983: An immature at Lough Gur from 23rd April to early May
        An immature at Aughinish, Limerick on 19th October
    1984: A juvenile at Charleville Lagoons on 25th August
    1985: One at Loop Head on 6th May
    1987: An immature at Charleville Lagoons on 20th June
    1988: A first year at Lough O'Donnell on 20th August
    1989: A first winter at Charleville Lagoons on 14th October
    1990: One at Aughinish, Clare on 6th January
        An immature at Fenloe Lake from 10th to 12th May

Sabines Gull* *Larus sabini*

Autumn passage migrant. Regular at the Bridges of Ross where, with the exception of an adult on 31st July 1988, all records refer to the autumn months. These are summarised as follows:

	August		September		October		**Total**	
	Adult	Juv	Adult	Juv.	Adult	Juv.	Adult	Juv.
1983	-	-	7	-	1	-	8	-
1984	-	-	2	27	-	5	2	32
1985	9	1	-	7	-	5	9	13
1986	-	-	-	-	-	1	-	1
1987	-	-	1	41	-	-	1	41
1988	-	-	4	68	1	1	5	69
1989	-	-	-	-	-	1	-	1
1990	2	-	16	9	1	1	19	10
1991	-	-	10	9	-	- -	10	9**
Total	11	1	40	161	3	14	54	176

Adult birds tend to occur earlier in the autumn than their juvenile counterparts with September accounting for the vast majority of birds.

Away from the Bridges of Ross, records were as follows;
  1985: A juvenile at Black Head on 7th October (G.D'Arcy)
  1987: A juvenile at Lough O'Donnell on 13th September (K.Grace)
      A juvenile inland at Charleville Lagoons from 14th to 16th September
      (T.Mee)
  1988: Juveniles at Horse Island and Kilbaha on 1st October (P.Brennan & T.Mee)

Black-headed Gull *Larus ridibundus*

Summer resident, widespread passage migrant and numerous winter visitor. A breeding survey of Lough Derg (Clare section) produced a total of 1331 nests from 23 colonies in 1985, and 785 nests from 22 colonies in 1987. The apparent decline between 1985 and 1987 may have been influenced by a number of factors including weather and fluctuations in water levels (J. Reynolds 1990). A total of 21 nests with 60 to 70 adults was present on East Sand Islet, Shannon Estuary on 21st April 1983. Significant colonies exist at Lough Bunny and Lough Cullaunyheeda. Some winter peaks include 1250 at Limerick City on 27th March 1985, 1000 at Clonderlaw Bay on 20th November 1982, 1000 at Lough Gur on 30th November 1986 and 1000 at Ballyvaughan on 10th January 1990.

36

Ring-billed Gull *Larus delawarensis*

Rare nearctic vagrant, mainly in winter. Regularly recorded at favoured sites in Limerick City such as Westfields, Thomond Weir and Clancy's Strand. Winter numbers are summarised as follows:

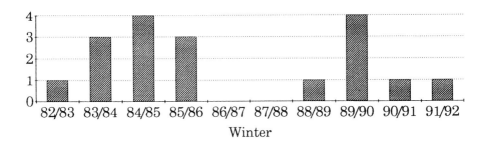

Away from Limerick City, other records include an adult and second-winter at Ballyallia Lake on 5th February 1985, a second-summer at Ballyvaughan on 20th May 1990, and adults at Ballyvaughan on 14th December 1991 and at Shannon Town on 23rd November 1991. There were no records for the region of this nearctic species prior to 1982.

Ring-billed Gull *(Paul Archer)*

Common Gull *Larus canus*

Very scarce resident, more numerous winter visitor and passage migrant. Breeds in very small numbers in north Clare and on Lough Derg. Widespread in winter throughout the Shannon Estuary, including Limerick City, in reasonable numbers and more numerous along the Clare coast. Some notable counts include 170 at Cliffs of Moher on 24th January 1982, 170 at Black Head on 29th January 1982, 179 on a Shannon Estuary count on 30th December 1984 and 150 at Seafield on 27th January 1990.

Lesser Black-backed Gull *Larus fuscus*

Scarce resident, more numerous passage and winter visitor. Breeds in small numbers on Mutton Island and possibly on some of the mainland sea-cliffs. More notable as a passage migrant particularly in Limerick. At Charleville Lagoons birds begin to arrive in early June with the main influx occurring during July and August. The lagoons are mainly used as a roost site with the freshly cut silage fields of the Golden Vale providing a rich food source. High autumn counts here include 500 on 22nd August 1986 and 450 on 28th August 1991. In recent years greater numbers are remaining to overwinter throughout the region with some notable peaks including 160 at Charleville Lagoons on 18th February 1990, 130 at Black Head on 29th January 1982, 90 at Limerick City dump on 27th March 1989 and 80 in Limerick City on 31st December 1991.

Herring Gull *Larus argentatus*

Resident and winter visitor. Although still a widespread breeder at numerous colonies on the Clare coast, this species has experienced a sharp decline in breeding numbers, matching a similiar nationwide downward trend. Only 64 pairs were counted at the Cliffs of Moher in 1987 compared to 770 in 1980. Also breeds in mixed colonies along with Great Black-backed Gulls at West Sand and West Dryanagh Islands in the Shannon Estuary. On 21st April 1983, 80 to 90 adults were present on the former and 60 to 70 adults at the latter island, and a few nests were found. Significant winter counts include 634 at the Cliffs of Moher on 24th January 1982, and 136 at Limerick City dump on 25th December 1982. A yellow-legged bird, probably of the American race *L. a. smithsonianus* was present at Kilkee on 13th April 1991.

Iceland Gull *Larus glaucoides*
Rare winter visitor. Annual totals are summarised as follows:

	1982	1983	1984	1985	1986	1987	1988	1989	1990	1991
Clare	-	2	2	1	1	2	1	-	2	2
Limerick	2	3	7	-	-	-	-	-	-	1

Only two of these records are adult birds. Records were from along the Clare coast, inland at Ennis and Limerick City dumps with a couple of records from the Shannon Estuary.

Glaucous Gull *Larus hyperboreus*
Rare winter visitor. Annual totals are summarised as follows;

	1982	1983	1984	1985	1986	1987	1988	1989	1990	1991
Clare	1	4	2	2	1	3	1	2	3	4
Limerick	-	1	1	1	1	-	-	-	-	2

The majority of these records refer to first winter birds. All the Clare records originate from along the coastline while the Limerick records all refer to the city area.

Great Black-backed Gull *Larus marinus*
Resident and winter visitor. Breeds in a number of colonies on the coast and on a number of islands in the Shannon Estuary. Some notable winter counts include 128 at the Bridges of Ross on 6th January 1983 and 120 at Ballyallia Lake on 30th January 1982.

Ross's Gull* *Rhodostethia rosea*
Rare arctic vagrant.
> 1988: An adult at Aughinish Causeway, North Clare on 28th February (T. Kilbane)

This is the first record of this species for the region.

Kittiwake *Rissa tridactyla*
Summer resident and passage migrant along the coast. There are a number of colonies along the Clare coast with the principal site being the Cliffs of Moher where 4038 nests were counted in June 1987. Also in June 1987, 690 nests were counted at Loop Head. Off the Bridges of Ross there is a regular spring and autumn passage, where peak hourly passage rates were 1233 flying south on 5th October 1983 and 1200 flying south on 7th October 1987. Also recorded occasionally in the Shannon Estuary on spring passage with notable counts of 40 at Beagh Castle on 17th May 1982 and 100+ at Tarbert Bay on 22nd May 1982.

Gull-billed Tern* *Gelochelidon nilotica*
Rare continental European vagrant.
1984: One at the Bridges of Ross on 21st September (P.Buckley)

This is only the third Irish record and the first for the region of this species.

Sandwich Tern *Sterna sandvicensis*
Summer visitor and passage migrant. Regular breeding colonies at Sturamus Island, Foynes and at Corranroo, North Burren where 30 pairs were recorded breeding in the former and 80 pairs in the latter site in 1989. On spring passage, birds are regularly recorded in the Limerick City area, while 25 at Liscannor on 30th March 1991 was a notable early count. Regular along the Clare coast in autumn with a peak count of 151 at Rine Point, Ballyvaughan on 6th August 1982.

Roseate Tern *Sterna dougallii*
Rare passage migrant. Only one record of a pair off the Bridges of Ross on 8th October 1987. This species is almost certainly under-recorded.

Common Tern *Sterna hirundo*
Summer visitor and passage migrant. Regular breeding colonies at Sturamus Island, Foynes and at Ballyvaughan Bay. In 1989, 30 pairs were recorded as breeding on Sturamus Island, 40 pairs at Gull Island, Ballyvaughan and 50 pairs at Carranroo Bay. One pair nested on the Cormorant Islands, Lough Derg in 1985 and two pairs were observed feeding young there on 23rd July 1987. Also one pair bred at Lough Cullaunyheeda in 1984. Regularly recorded throughout the region during spring and autumn passage. An unusual occurrence was a late bird inland at Lough Gur on 20th October 1984 following severe gales.

Arctic Tern *Sterna paradisaea*
Summer visitor and passage migrant. In 1984, six breeding pairs were recorded at an estuarine colony in Clare (Tern Survey). In 1989 three pairs bred at Gull Island, Ballyvaughan Bay and two pairs were recorded as breeding at Sturamus Island, Foynes. Regularly recorded along the Clare coast on autumn passage. During autumn coverage at the Bridges of Ross, 223 were recorded in 1990 with a daily peak of 76 on 19th September 1990.

Little Tern *Sterna albifrons*

Scarce passage migrant. The only spring record was a pair at Liscannor on 6th May 1990. Two were present off Beagh Castle (with Black Terns) on 30th September 1982. The following records all refer to the Bridges of Ross:

1984: Two on 19th September and one on 19th October
1985: One on 15th August and two on 6th and 14th September
1988: One on 13th October
1989: Six on 26th August

Black Tern *Chlidonias niger*

Scarce passage migrant. The only spring records come from Limerick City where single birds were seen on 22nd May 1987 and again during the month of May 1988.

Autumn coverage at the Bridges of Ross produced the following records:

1983: Six on 16th September and one on 5th October
1987: Two on 16th September
1988: Two on 27th August and one on 6th September
1989: One on 7th October
1990: One on 15th August and two on 27th October
1991: One on 15th September, two on 21st September and one on 22nd September

Autumn records from elsewhere were as follows:

1982: Two at Glin on 27th September
Four off Beagh Castle on 30th September and six there on 4th October, two on 7th October with one remaining until 8th October
1983: Two were seen on Lough Derg in early October.

Interestingly, the 1982 Beagh Castle birds were accompanied by Little Terns.

White-winged Black Tern* *Chlidonias leucopterus*

Rare passage migrant.

1984: An adult at the Bridges of Ross on 19th September (P.Buckley et al).
1990: A juvenile at Charleville Lagoons on 28th October (T.Mee et al).

These are the second records for Clare and Limerick respectively.

Guillemot *Uria aalge*

Summer resident and passage migrant with a small number remaining to overwinter. Scattered colonies exist along the Clare coast with the principal site being the Cliffs of Moher, where in June 1987, 12207 individuals were counted. This shows only a marginal decrease from the 1980 figures of 12882. On 24th May 1987, 4010 individuals were counted at the Loop Head colony. Regularly recorded during seawatch coverage at the Bridges of Ross where peak hourly counts include 1442 on 18th February 1982 and 1335 on 5th October 1983.

Razorbill *Alca torda*

Summer resident and passage migrant. Colony distribution mirrors that of Guillemot, but in much smaller numbers. At the Cliffs of Moher, 2188 individuals were counted in June 1987. When compared to the 1980 figure of 2826, an apparent decline of 22.5% has occurred at the Cliffs of Moher. On 24th May 1987, 105 individuals were counted at the Loop Head colony with 33 definite nest sites found. A juvenile was found dead at Corbally, Limerick on 7th September 1989.

Black Guillemot *Cepphus grylle*

Resident. Breeds in small numbers scattered along the Clare coast and probably along the outer Shannon Estuary from Carrigaholt to Loop Head. At Mutton Island 14 birds were recorded on 19th May 1990. Other notable counts were 20 at Carrigaholt on 16th July 1983, 14 at the Rine, Ballyvaughan on 27th January 1990 and 10 at Black Head on 25th April 1991.

Black Guillemot *(Paul Archer)*

Little Auk* *Alle alle*

Autumn and winter passage migrant. All records refer to the Bridges of Ross unless otherwise stated:

1982: One on 18th October (P.Buckley)
1984: A total of 71 between 10th September and 23rd November, with a maximum of 30 on 23rd September (P.Buckley et al)
1985: One on 18th March (P.Buckley)
One on 22nd October (K.Mullarney)
1988: One at Aughinish, North Clare on 17th January (T.Kilbane et al)
One found dying at Broadford, East Clare on 2nd February (J.Reynolds)
1989: One on 7th October (T.Mee et al)
1991: One found dead at Aughinish Point, North Clare on 26th January (R.Ackerly)
31 on 21st and 337 on 22nd December (A.Kelly et al)

The exceptional December 1991 passage at the Bridges of Ross illustrates the greater potential for recording this species during winter, rather than autumn, seawatches.

Puffin *Fratercula arctica*

Summer resident and passage migrant. In 1987, 1335 individuals were counted at the Cliffs of Moher. This is almost identical to the 1980 figure of 1355. Probably underrecorded on passage at the Bridges of Ross, where the peak autumn total was 136 in 1985.

Rock Dove *Coloumba livia*

Resident along the west Clare coast. Some notable counts include 17 at Illaunaeraun on 28th December 1982, 22 at Loop Head on 15th October 1988, and 25 at Martello Tower, Aughinish on 28th January 1990.

Stock Dove *Columba oenas*

Resident. Relatively scarce in most parts of Clare. Some notable counts include 40 at Mungret on 25th November 1984 and 10 at Sixmilebridge on 24th March 1991.

Woodpigeon *Columba palumbus*

Very numerous and widespread resident. An exceptional count was a flock of 600 over Ballyallia Lake on 11th December 1990.

Collared Dove *Streptopelia decaocto*

Widespread resident. Absent by and large from the dairy heartland of south-east Limerick. Counts include 14 at Ennis 17th November 1982 and 10 at Kilbaha on 6th May 1988.

Turtle Dove *Streptopelia turtur*

Scarce summer visitor and passage migrant.

    1982: Two at Adare on 4th June
    1987: One at Adare on 10th May. Single birds at Loop Head on 19th and 29th
           September. A juvenile at Ballyorgan, Limerick on 3rd October
    1988: Single birds at Loop Head on 16th and 19th September
    1989: Single birds at Loop Head on 30th September, 1st and 4th October. These
           records probably refer to the same bird.
    1990: A juvenile at Shannon on 24th July

Cuckoo *Cuculus canorus*

Summer visitor. Mainly confined to upland areas and lowland reedbed habitat in both counties. Earliest report was one at Sixmilebridge on 15th April 1989. The only multiple sightings were three males at Glenanaar, Ballyhouras on 5th May 1986, and two at Fenloe Lake on 6th May 1990.

Barn Owl *Tyto alba*

Resident. Thinly distributed throughout the region.

Long-eared Owl *Asio otus*

Resident. Almost certainly widespread throughout both counties, but the only reports were from Cratloe, Fenloe Lake, Lough Gur, Shannon Airport Lagoon and Glenroe, Limerick.

Short-eared Owl *Asio flammeus*

Scarce summer visitor, passage migrant and winter visitor. Almost certainly bred in Limerick during the period. A single bird was seen at an undisclosed site on 11th April 1982 and again on four occasions from 25th May to 30th June 1983. Singles were seen in mid July 1986 on Woodcock Hill and at Patrickswell from 29th July to 4th August 1986.

Autumn and winter sightings were as follows:

    1982: One found dead at Shannon Airport Lagoon on 28th October. Singles at
           Shannon from 1st November to 22nd December
    1983: One at Westfields on 31st January and one at Ing on 4th December. One to
           two wintered on Aughinish Island, Limerick during 1982/83
    1985: One at Limerick City on 3rd November
    1988: Singles at Loop Head on 21st September and 16th October
    1990: Singles at Clenagh on 6th January and at Loop Head on 25th October
    1991: Singles at Loop Head on 12th October and at Shannon on 23rd October

Nightjar *Caprimulgus europaeus*

Rare summer visitor.

    1984: A female at Clooncoose, south Burren in July.
    1985: One near Bell Harbour from 6th to 13th July.
    1986: One near Bell Harbour from 28th June to 1st July.

Swift *Apus apus*

Widespread summer visitor. Breeding in most towns and villages. A large concentration of 1500 feeding low over Charleville lagoons on 29th July 1986 was an exceptional count. This species is also very much at home at higher altitudes as illustrated by a party of 20 seen around the summit of Carron mountain, Ballyhouras (c.1600 ft) on 14th June 1986. A very early arrival was one at Castletroy on 12th April 1991.

Kingfisher *Alcedo atthis*

Resident. Does not favour the faster flowing upland streams and rivers in the region. Particularly numerous along the River Shannon between Limerick and Killaloe.

Belted Kingfisher* *Ceryle alcyon*

Rare nearctic vagrant.

> 1984: One at Ballyvaughan from 28th October to early December 1984 (G.D'Arcy et al). Presumably the same bird, a female, was relocated on the Tipperary side of Killaloe from 6th February to 21st March 1985.

This was the third Irish, and fifth Western Paleartic record of this species, and a first for the region. It also caused the arrival of an equally rare species for the region, i.e. the British 'twitcher'!

Bee-eater* *Merops apiaster*

Rare vagrant.

> 1990: One heard at Black Head on 27th May flying north (C. Murphy)**.

If accepted, this will be the first record for the region of this southern European species.

Great Spotted Woodpecker* *Dendrocopos major*

Rare vagrant.

> 1990: A single bird was sighted at the Rine, Ballyvaughan on 2nd May (S.C. Hutchings)**

This species was last recorded in Clare in 1968.

Skylark *Alauda arvensis*

Summer resident, passage migrant and winter visitor. Mainly confined to upland moorland, amenity areas, roadside verges and coastal areas in summer. In winter, birds tend to vacate the uplands and small flocks can be encountered in more open lowland terrain. At Loop Head some interesting movements have occurred in both spring and autumn, with notable counts of 121 on 23rd April 1984 and 206 on 21st September 1988.

Sand Martin *Riparia riparia*

Widespread summer visitor. Almost completely dependent on worked sandpits for breeding colony sites. Coastal sand dunes in west Clare are also utilised, but natural river sand-banks no longer seem to feature, probably due to extensive arterial drainage throughout the region. Earliest date is one at Ballyallia Lake on 11th March 1983. Some passage counts include 1000 at Meelick Bay, Lough Derg on 29th July 1985, 500 at Charleville Lagoons on 11th August 1984 and 500 at Meelick Bay, migrating south, on 20th August 1983.

Swallow *Hirundo rustica*

Widespread and common summer visitor. Some interesting passage counts include 1000 roosting in reedbeds at Shannon Airport Lagoon on 6th September 1983, 150 roosting on nettles at Charleville Lagoons on 3rd May 1986, 250 at Westfields on 18th September 1983 and 300 at Loop Head on 9th September 1990.

Earliest date: Two at Westfields on 28th March 1982.
Latest date: One at Cratloe on 10th November 1990.

Red-rumped Swallow* *Hirundo daurica*

Rare vagrant.

1987: One at Loop Head on 7th November (P.Brennan, P.Buckley,T.Mee, T.Tarpey, et al)

This is the first record for the region of this extremely rare southern European species.

House Martin *Delichon urbica*

Widespread and relatively common summer visitor. Though numbers tend to fluctuate from year to year, this species has certainly benefited from the 'bungalow boom'. Sizable colonies exist at Mullagh Church, west Clare and at Ennis. Very scarce on migration at Loop Head with one exceptionally late record on 26th October.

Richard's Pipit* *Anthus novaeseelandiae*

Rare vagrant.

1987: One at Loop Head on 6th and 7th October (P.Brennan, J.H.Grant, D.Manley et al)
1990: One at Loop Head on 16th October** (K.Preston)

These are the first and second records for the region of this rare Siberian species.

Tree Pipit* *Anthus trivialis*

Rare passage migrant.

1985: One trapped at Kilbaha on 20th October (P.Brennan et al)
1991: One at Loop Head on 2nd August (J.Murphy)**

Meadow Pipit *Anthus pratensis*

Widespread and common resident. Largely absent from farmland except in rough or unimproved pasture in summer. In winter, upland moorland populations mostly move to more lowland areas. Notable autumn concentrations occur at Loop Head with counts of 500 on 4th October 1987 and 9th September 1990.

Rock Pipit *Anthus spinoletta*

Widespread and common resident along the Clare coast and also along the Shannon Estuary. Forty-three birds were seen at Clahane, Liscannor on 5th February 1987 and up to 100 have been recorded at Loop Head on a number of occasions. A nest was found at Dromoland in May 1983, at a considerable distance from the Fergus Estuary.

Yellow Wagtail *Motacilla flava flavissima*

Scarce passage migrant.

1988: One at Loop Head on 2nd October
1989: One at Loop Head on 30th September

These are the first and second records for the region of this species.

Grey Wagtail *Motacilla cinerea*

Widespread and common resident of wetland habitat, largely along freshwater streams and rivers.

Pied Wagtail *Motacilla alba yarelli*

Widespread and common resident. The only information received was a peak of 860 birds in March 1985 roosting at the Regional Hosptial at Dooradoyle, Limerick.

Birds of the nominate race White Wagtail *M. a. alba.*

Passage migrant in spring and autumn. Some peak spring counts include 30 at Seafield on 20th April 1982 and 25 at Aughinish, Limerick on 29th April 1983. Peak autumn counts were 40 at Shannon Airport Lagoon on 1st September 1982 and 20 at Kilbaha on 13th September 1986.

Waxwing* *Bombycilla garrulus*

Rare winter visitor.

1988/89:	One at Shannon Town from 13th December 1988 to 10th January 1989 when unfortunately it was found dead (P.Brennan et al).
1989:	Three at Shannon Town on 27th January 1989 (A.Flynn et al)**
1990:	One at Shannon Town from 17th February to March(A.Flynn et al) **

This series of records is the first for Clare since the 1965 invasion.

Waxwing *(Allan Mee)*

Dipper *Cinclus cinclus*
Resident. Present on most suitable shallow and fast-running streams. Only significant count was nine at Broadford, Clare on 17th January 1990.

Wren *Troglodytes troglodytes*
Resident. Widespread and very common. In a breeding birds census of five farmland plots in the vicinity of Limerick City, the Wren proved by far the most abundant breeding species, accounting for 22% of the total breeding population.

Dunnock *Prunella modularis*
Resident. Widespread and common. In a breeding birds census of five farmland plots in the vicinity of Limerick City, the Dunnock accounted for 11% of the total breeding population.

Robin *Erithacus rubecula*

Resident. Widespread and common. In a breeding birds census of five farmland plots in the vicinity of Limerick City, the Robin accounted for 12% of the total breeding population.

Bluethroat* *Luscinia svecica*

Rare continental vagrant.

> 1982: A male of the red-spotted race *L. s. svecica* was present at Shannon Airport Lagoon on 6th and 7th October (P.Brennan et al)

This is the first record for the region of this species.

Black Redstart *Phoenicurus ochruros*

Scarce autumn passage migrant. Occasionally overwinters.

> 1982: A pair at Foynes Island on 11th January
> One at Robertstown on 2nd December and two at Aughinish, Limerick on 22nd December
> 1983: Three to four at Shannon Airport Lagoon from 4th January to 17th March
> One at Shannon Town from 26th January to 16th February
> One at Cratloe on 30th January
> One at Quilty from 11th January to 17th March
> One at Shannon on 21st November
> 1985: One at Loop Head on 19th and 20th October and four there on 21st October
> 1986: One at Shannon Airport Lagoon from 19th to 25th January
> 1987: Two at Loop Head on 1st November and singles there on 2nd and 4th November
> 1988: At Loop Head, singles were present on four dates between 22nd October and 12th November. Also two were present on 29th and 30th October and four on 6th November
> 1989: One at Shannon Airport Lagoon in November
> 1990: Singles at Loop Head on 25th and 26 October. Three on 19th and one on 30th November at Shannon.
> 1991: Two at Loop Head on 12th October

The favourite haunt of this species at Loop Head is in and around the lighthouse buildings.

Redstart *Phoenicurus phoenicurus*

Passage migrant in autumn.

> 1985: A first year male at Kilbaha on 12th October
> 1987: Singles at Kilbaha on 5th and 10th October
> 1988: A female on 20th September and a juvenile male on 16th October at Loop Head
> 1989: Singles on 1st October, 7th and 8th October and two on 4th October at Loop Head
> 1990: One at Loop Head on 21st October
> 1991: Singles at Loop Head on 10th, 12th and 13th October

These are the first records for the region of this species.

Whinchat *Saxicola rubetra*

Rare summer visitor and passage migrant. A female and juveniles were seen at a site near Carrigkerry, west Limerick during June and July 1989. This constitutes the first ever breeding record for this species in Limerick. The only other summer record is of a female at Ballyvaughan on 3rd June 1983.

Autumn records all refer to the Loop Head coverage area with details as follows:

1983: Five on 1st October
1986: One on 6th September
1987: Five on 4th October
1988: One on 22nd and 23rd, three on 26th and one on 28th October
1989: One on 4th October
1991: One on 12th October

Stonechat *Saxicola torquata*

Widespread and common resident along the Clare coast and on both sides of the Shannon Estuary. Also smaller populations occur inland at most freshwater lakes and in upland moorland/young forestry plantations. Very numerous at Loop Head during autumn coverage with a peak count of 50 on 1st October 1989.

Wheatear *Oenanthe oenanthe*

Localised summer visitor and passage migrant. Breeds mainly near the Clare coast and to a lesser extent along the Shannon Estuary shore. Earliest report of spring arrival was two at Kilkee on 9th March 1991. Most notable spring count was of 15 at Loop Head on 23rd April 1984 and 10 at Loop Head on 13th October 1991, with the latest sighting being one at Loop Head on 27th October 1991. Occasionally occurs inland in Limerick during autumn passage, with reports from Charleville Lagoons and Lough Gur.

Wheatear *(John Murphy)*

Pied Wheatear* *Oenanthe pleschanka*

Rare eastern European vagrant.

1988: A first winter male at Kilbaha from 5th to 8th November (P.Brennan,
T.Mee, T.Tarpey et al)

This is only the third Irish record and the first for the region of this species.

Gray-cheeked Thrush* *Catharus minimus*

Rare nearctic vagrant.

1991: A juvenile was trapped at Kilbaha on 12th October (P.Brennan et al)**

This is only the fourth Irish record and the first outside of Cape Clear.

Ring Ouzel *Turdus torquatus*

Rare passage migrant.

1987: One at Loop Head on 7th November

Unfortunately, despite an increase in observers and coverage during the 1980s there continues to be very few records of this species.

Blackbird *Turdus merula*

Widespread and common resident and winter visitor. A perceptible increase in numbers occur at Loop Head in late October, coinciding with the arrival of other wintering thrushes. A bird weighing only 77 grams (normal weight c. 90g) on 1st November 1988 was an obvious migrant with depleted fat reserves. A total of 55 was recorded at Loop Head on 7th November 1987.

Fieldfare *Turdus pilaris*

Widespread and common winter visitor. Earliest autumn arrival date at Loop Head was 9th October. Only significant counts were 200 at Ballyorgan on 7th December 1985 and 120 at Loop Head on 13th November 1989. The latest spring sighting was one at Ballyorgan on 25th April 1987.

Song Thrush *Turdus philomelos*

Widespread and common resident and winter visitor. At Loop Head, the arrival of migrants of this species in late October and early November is quite apparent when they frequently outnumber other thrush species. Some peak counts at Loop Head include 140 on 7th November 1987 and 125 on 28th October 1988. A bird ringed at Kilbaha on 27th October 1987 was found dead at O'Briens Bridge, east Clare on 10th March 1988.

Redwing *Turdus iliacus*
Widespread and common winter visitor. Earliest arrival date at Loop Head was two on 30th September 1991. Some notable peak counts include 800 at Ballyorgan on 22nd December 1984, 180 at Loop Head on 31st October 1987 and up to 150 roosting in alders at Westfields during the 1982/83 winter. Latest spring sightings were two at Fenloe on 1st April 1989, and an unexpected summer record was a male (in song) trapped at Shannon Airport Lagoon on 28th June 1991. This is the first June record of this species in Ireland.

Mistle Thrush *Turdus viscivorus*
Widespread resident but considerably less numerous than the other common thrush species. One notable count from Loop Head involved 11 birds on 24th October 1991.

Grasshopper Warbler *Locustella naevia*
Localised summer visitor. Particularly numerous in damp lowland areas such as around Shannon Town and Limerick City. Also widespread in the boggier areas of west Clare where three males were heard at Loop Head in June 1988. Also occurs in upland areas particularly in young forestry plantations with reports of three birds calling at Gallows Hill (700ft) in late July, 1986 and birds regularly heard most years at Glenanaar, Ballyhouras (900ft). Earliest spring date was one at Shannon on 16th April 1989.

Sedge Warbler *Acrocephalus schoenobaenus*
Locally common summer visitor. Breeds in good numbers in almost every wetland marsh throughout the region. At Westfields in 1983, birds were recorded from 26th April to 29th September with a minimum breeding total of seven pairs. The birds favoured an extensive area of sedge, but all nests were situated within five metres of willow fringe.

In autumn, birds are more exclusively found in reedbed habitat where large hatches of aphids provide an abundant food source and a means of putting on much needed pre-migratory fat. Large concentrations, particularly of juvenile birds, occur at reedbed sites along the Shannon Estuary such as at Shannon Airport Lagoons during late July and early August. The peak Sedge Warbler season at Shannon Airport Lagoon was 1991 when a total of 673 birds was ringed.

Reed Warbler *Acrocephalus scirpaceus*

Rare autumn passage migrant.

> 1983: One trapped at Shannon Airport Lagoon on 3rd October
> 1985: Singles trapped at Shannon Airport Lagoon on 26th September and 14th October
> 1986: One at Kilbaha on 1st October
> 1990: One at Kilbaha on 16th October
> 1991: One at Kilbaha on 7th September

Still no signs of any pioneering movements into the region from the now well established south coast population.

Melodious Warbler* *Hippolais polyglotta*

Rare autumn migrant.

> 1991: One at Kilbaha on 28th August (D.McAdam)**

This is the first record for the region of this species.

Barred Warbler* *Sylvia nisoria*

Rare autumn migrant.

> 1985: A juvenile was trapped at Kilbaha on 20th October (P.Brennan et al)

This is the first record for the region of this species

Lesser Whitethroat *Sylvia curruca*

Rare autumn migrant.

> 1990: One was trapped at Kilbaha on 16th October (T.Mee)

This is the first record for the region of this species.

Whitethroat *Sylvia communis*

Fairly widespread summer visitor. Particularly numerous in the drier limestone areas such as the Burren and in the Barrigone area of west Limerick, where six singing males were calling on Aughinish Island during May 1983. A very late bird was present at Loop Head on 28th October 1987.

Garden Warbler *Sylvia borin*

Rare summer visitor and passage migrant. A singing male was seen on Red Island, Lough Derg on 21th July 1987, and a pair bred at Kilmacduagh, north Clare in 1989. An exceptional winter record is of a bird frequenting a bird table on a number of occasions on 12th February 1983 at Crescent College Comprehensive, Limerick. This is only the third known winter record for Ireland of this species.

Regularly recorded in the Loop Head area during autumn coverage with details as follows:

1985: Singles on 12th and 20th October
1986: One on 4th and 5th October
1987: A total of 10, with two on 31st August and the rest in October, with the last on 24th October
1988: A total of seven. Recorded from 17th September until 17th October, with two on 14th October
1989: One on 1st October
1990: Singles on 15th September and 30th October and a corpse found at the light on 21st October
1991: Singles on 8th September and 10th October

Blackcap *Sylvia atricapilla*

Scarce summer visitor and more numerous winter visitor. From 1983 to 1986 a total of 29 birds was heard over the four summers at Curraghchase. The only female seen was in 1985, when one was observed carrying nest material.

During a breeding birds census on a 118 acre plot at Cratloe, three territories were identified in 1986. Other reports of singing males include two at Dromoland on 30th April 1984, one at Inchicrowan on 7th May 1989, at least three at Dromore Wood on 1st June 1991 and three at Broadford, Clare on 9th June 1991. Breeding was also reported from Doon Lough and Lough Gur.

Winter records are too numerous to mention in detail, but to summarise, birds were recorded throughout the region even in some unlikely spots like on a rock beach at Clahane, Liscannor. An exceptional total of 20 was estimated as wintering in the Shannon Town area during January, 1990. Up to four birds have been recorded at certain well-watched Limerick City gardens.

Autumn passage at Loop Head begins as early as 15th September with most occurring in October and a maximum of six on 14th November 1989. Loop Head autumn totals summarised as follows:

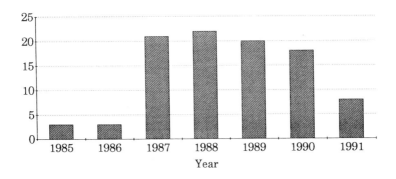

Arctic Warbler* *Phylloscopus borealis*

Extremely rare autumn migrant.
>1986: A single bird at Kilbaha on 6th September (P.Brennan, J.Murphy)

This is only the sixth Irish record of this species.

Yellow-browed Warbler* *Phylloscopus inornatus*

Rare Siberian vagrant.   Recorded on autumn passage almost annually at Kilbaha.
>1985: Individuals at Kilbaha on 12th and 19th October and two at Kilbaha on 21st October (P. Brennan et al)
>1986: One at Kilbaha on 5th October (P.Brennan et al)
>1988: Singles at Kilbaha on 16th, 17th and 26th October (P.Brennan et al)
>1990: Singles at Kilbaha on 14th, 16th and 20th October and 5th November (P.Brennan et al)
>1991: One at Kilbaha on 26th October (P.Brennan et al)

These records are the first for the region of this species.

Chiffchaff *Phylloscopus collybita*

Common breeding summer visitor, wherever there is mature woodland.
Each year a few birds have been recorded during the winter in Limerick City at Westfields, in Ennis and at a few other localities in both counties. An autumn passage movement is evident at Loop Head in October and November of most years.

Records of the Siberian race *P. c. tristis*
>1990: One was present at Loop Head from 4th to the 14th November

Willow Warbler *Phylloscopus trochilus*

Very common breeding species.   An autumn passage movement is evident at Loop Head each year, with greatest numbers seen in September and early October, when peak counts of up to 20 in a day are recorded.   The latest record was one on 3rd November 1990.

Goldcrest *Regulus regulus*

Common resident and summer visitor. In most years a definite autumn passage is evident at Loop Head in late September and October, with peaks of up to 60 birds in a day.

Firecrest* *Regulus ignicapillus*

Rare autumn migrant.
>1989: A male, trapped at Killula, Newmarket-on-Fergus on 25th November (P.Brennan)**

This is only the second record of this species which surprisingly, has not yet been recorded at Loop Head, much to the chagrin of local birders!

Spotted Flycatcher *Muscicapa striata*
Summer visitor. A notable late record is of a single bird at Kilbaha which remained from 27th October to 14th November 1987.

Red-breasted Flycatcher* *Ficedula parva*
Rare autumn migrant. Five records of single birds at Kilbaha/Loop Head. Four of the sighting were in October with the remaining one in September.
   1985: Singles at Kilbaha on 6th and 21st October (P. Buckley)
   1988: First-winter bird trapped at Kilbaha on 12th October (P.Brennan et al)
   1989: One trapped at Kilbaha on 30th September (T.D. Carruthers)**
   1991: One trapped at Kilbaha on 10th October (D. Barton)**

These records are the first for the region of this species

Pied Flycatcher *Ficedula hypoleuca*
Regular passage migrant at Loop Head/Kilbaha, having been recorded in September or October every year since 1985. A single bird was present at Quilty from 10th to 13th March 1990. This is the earliest spring record ever in Ireland.

   Autumn records at Kilbaha/Loop Head were as follows:
   1985: Five between 6th and 13th October
   1986: Singles on 6th September and 4th/5th October
   1987: Singles on 20th September and 5th October
   1988: Singles on 10th, 16th and 20th September
   1989: One on 1st October
   1990: One from 16th to 25th October
   1991: Singles on 7th and 8th September

Long-tailed Tit *Aegithalos caudatus*
Common resident.

Coal Tit *Parus ater*
Common resident.
   1985: An unprecedented autumn irruption of this species throughout Britain
         and Ireland was reflected in a flock of 52 trapped at Kilbaha on 12th and
         13th October
   1991: Peak of 18 on 21st October at Loop Head

Blue Tit *Parus caeruleus*
Common resident. The only notable record was of 40 at Loop Head on 1st October 1989.

Great Tit *Parus major*
Common resident. The only notable record was of 10 at Loop Head on 30th September 1990.

Treecreeper *Certhia familiaris*
Common resident. The only record of note was of eight at Curraghcase on 20th November 1988.

Jay *Garrulus glandarius*
Resident, breeding in small numbers in a number of woodlands throughout Clare and Limerick. The most notable record was of eight at Cratloekeel on 10th January 1986. The only other record was of three birds at Darragh, Glenroe on 16th November 1983.

Magpie *Pica pica*
Common resident. The only record of note was 45 at Loop Head on 22nd October 1990.

Chough *Pyrrhocorax pyrrhocorax*
Chough populations have remained relatively stable over the last 30 years. The 1982 Chough survey recorded between 31 and 34 pairs breeding along the Clare coast and an additional 35-37 non-breeding birds, giving a total of approximately 100 birds. This figure is roughly in line with the population level found in the previous Chough survey in 1963. The sighting of a single bird at Shannon on 6th January 1991 was the only record away from the west coast. Up to 20 birds have been recorded almost annually at Loop Head in October

Jackdaw *Corvus monedula*
Common resident. There is an annual movement of flocks of Jackdaws moving south over Kilbaha towards Loop Head in October and early November. In one such movement, a flock of 70 birds was seen to disappear out to sea in a south-westerly direction off Loop Head on 14 October 1988.
> 1989: A peak of 500 birds on 23rd October at Loop Head
> 1990: A peak of 2000 birds on 19th October at Loop Head
> 1991: A peak of 1500 birds on 19th October at Loop Head

Rook *Corvus frugilegus*
Common resident. Often accompanies Jackdaws in autumnal flock movements at Loop Head. 500 were noted on 19th October 1990, during a period of active Jackdaw movements.

Hooded Crow *Corvus corone cornix*
Common resident. The only notable records were 30 at Loop Head on 26th October 1989 and 100 at Charleville Lagoon on 2nd June 1991.

Birds of the nominate race Carrion Crow *C. corone corone.*
There were two sightings of Carrion Crow, two at Slieve Barnagh on 29th
July 1987 and one at Loop Head on 29th September 1990.

Raven *Corvus corax*
Resident, commonly breeding in upland areas. The only records received
are of 17 birds on Slieve Barnagh on 2nd October 1982, 16 at
Newmarket-on-Fergus on 22nd March 1984, 15 birds on the Ballyhouras on
16th August 1984 and nine at Loop Head on 6th November 1991.

Starling *Sturnus vulgaris*
Very common resident and winter visitor. The only notable records are of
at least 2500 birds approaching a roost in a forestry plantation at
Ballinvreena, Kilfinane on 15th January 1987 and 1000 birds at Loop Head
on 15th September 1990. A bird showing evidence of partial albinoism
was sighted at Ballyorgan on 8th November 1983.

Rose-coloured Starling* *Sturnus roseus*
Rare continental vagrant.
> 1989: An adult at Shannon Town on 29th July (J. Murphy)

This is only the fourth Clare and regional record, and the first this century.

House Sparrow *Passer domesticus*
Common resident. The only notable record was 45 at Kilbaha on 11th
November 1990.

Tree Sparrow *Passer montanus*
Rare resident. Breeds annually in small numbers at Loop Head, and along
the Clare coast. At Loop Head, the peak count was 14 on 19th October
1985, and although there were no records in 1986, birds have been seen in
all subsequent years. The breeding range of this species has probably
contracted over the last few years.

Chaffinch *Fringilla coelebs*
Very common resident and winter visitor. Migrants arrive at Loop Head in
late October. The most spectacular movement was in 1988 when regular
falls occurred from mid October into November, peaking with a total of
2020 on 29th October. The majority of these birds was probably of the
Continental races *F. c. hortensis* and *F. c. coelebs*. Other large counts
include 500 at Castleoliver, Ballyorgan on 30th November 1985, 600 on
26th October 1989 and 250 on 19th October 1990 the latter two at Loop
Head. The only other record of note was an albino seen at Shannon on 29th
January 1989.

Brambling *Fringilla montifringilla*

Regular winter visitor, recorded in every year except 1983. The majority of records are of individual birds from a variety of localities including, Foynes, Castleoliver, Cratloe, Sixmilebridge, Castletroy and Curraghchase. Some birds are seen at Loop Head on passage in autumn, with the peak count being six on 2nd November 1987.

 1982: Four at Foynes on 14th January
 1984: A single male at Ennis on 4th February
 1985: A female at Castleoliver, Ballyorgan on 27th December
  A single bird at Loop Head on 27th October
 1986: A single bird at Cratloekeel, Clare on 19th February and five at
  Sixmilebridge on 29th November
 1987: Singles at Loop Head on 24th and 27th October, and 7th November
  Six present on 2nd and two on 14th November
  One at Castletroy on 15th January
 1988: Singles at Loop Head on 16th and 28th October, and 3 on 6th November.
  Two at Curraghchase on 11th November
 1989: Four on 26th October and singles on 13th and 15th November at Loop Head.
  One at Castletroy on 26th January
 1990: A male at Shannon on 4th February and a female at the same locality from
  1st to 5th March
 1991: One at Loop Head on 23rd October

Greenfinch *Carduelis chloris*

Common resident. A flock of between 900 and 1000 birds was seen perched on wires at Corranroo, Burren on 24th August 1991. Another large flock of 200 birds was seen at Castleoliver, Ballyorgan on 7th December 1985. In some years, a small passage is evident at Loop Head in October.

Goldfinch *Carduelis carduelis*

Common resident. The only notable count was 80 at Fenloe on 10th December 1990.

Siskin *Carduelis spinus*

Localised resident and winter visitor. There is some evidence of passage at Loop Head where birds have been recorded in fluctuating numbers, mainly in October. The largest count was 75 birds on 20th October 1988, which coincided with a large influx of Chaffinches. Outside Loop Head, 65 birds were seen at Ballyorgan on 14th November 1983 and 20 were recorded at Fenloe on 25th March 1990. From January to March 1991, a total of 36 birds was trapped and ringed in gardens in Shannon Town, which is well above the normal numbers attracted to garden feeders in that area in winter.

Linnet *Carduelis cannabina*
Common resident. The largest count was 400 at Sixmilebridge on 24th March 1989. Other high counts include 300 at Castleoliver, Ballyorgan on 7th December 1985 and 70 at Fenloe Lake on 1st August 1990. Flocks of up to 100 birds are seen regularly at Loop Head each October, the peak count being 250 on 6th October 1987.

Twite *Carduelis flavirostris*
Very scarce breeder and winter visitor. The *New Breeding Atlas 1988-91* proved breeding in the Liscannor/Hags Head area and surprisingly from the Limerick side of the Shannon Estuary. This latter record is the first confirmation of breeding for Limerick during the present century. The only other records are three at Shannon Airport Lagoon on 10th January 1982, five at Quilty on 6th January 1987, a single male at Loop Head on 22nd October 1988 and one at Bridges of Ross on 26th August 1989.

Redpoll *Carduelis flammea*
Resident. The only records received of this species are from winter, with the highest count being 72 at Ballyvaughan on 1st January 1982.

Birds of the northern continental European race, Mealy Redpoll *C. f. flammea* :
> 1986: A flock of 25 birds in the Kilbaha area on 31th October with at least six
> birds remaining until 1st November.

Crossbill *Loxia curvirostra*
Prior to 1987 there was only one record for this species in the region, a single bird at Loop Head on 20th October 1986. Since then small flocks of up to 20 birds have been reported from a number of localities throughout the region, and breeding was confirmed for Coolfree Mountain, Ballyhouras in 1987.

The records received are as follows;
> 1986: One at Loop Head on 20 October
> 1987: Seven in April and 20 in May in Limerick. A flock of 20 birds at Coolfree
> Mountain, Ballyhouras included at least six juveniles on 16 th May
> 1989: 19 at Dromore Wood on 2nd February
> 1990: 20 at Loop Head on 13th October and two at Shannon on 10th November
> 1991: Six at Cratloe on 24th March
> A flock of 30 at Blackrock, Ballyhouras in June.
> Six, including a first year male and female at Glenanaar, Ballyhouras on
> 24th December

Scarlet Rosefinch* *Carpodacus erythrinus*

Rare European migrant.

> 1987: One bird, either a female or immature, at Kilbaha on 6th October (J.H. Grant, S.H. Piotrowski, C.P.S. Raffles et al).

This is the first record for the region of this species.

Bullfinch *Pyrrhula pyrrhula*

Common resident.

Hawfinch* *Coccothraustes coccothraustes*

Rare European vagrant.

> 1988/89: About 35 birds, increasing to 95 by February and March, were present at Curraghcase Forest Park from 30th October 1988 to 25th March 1989 (T.Tarpey, A. Walsh, K. Grace, et al). This flock is the largest gathering ever recorded in Ireland and was the highlight of an influx of this species into Ireland during the autumn of 1988.
>
> 1991: Two adults and a juvenile at Ballyvaughan on 3rd and 4th September (B. Clark, J. Clark). There is a distinct possibility that these birds bred in Ireland in 1991, perhaps having remained here from the autumn influx of 1988.

These are the first records of this species for the region.

Hawfinch *(Tom Tarpey)*

Yellow-rumped Warbler* *Dendroica coronata*

Rare nearctic vagrant.

> 1986: A first winter bird trapped at Kilbaha on 31st October remaining until 2nd November (P.Brennan et al).

This is the first regional and sixth Irish record, and the first away from Cape Clear.

Lapland Bunting *Calcarinus lapponicus*
Rare autumn migrant. All the sightings of this species eminate from Loop Head, where one or two birds were recorded annually from 1986 to 1990.

1985: One   on 26th October
1986: One   on 4th October
1987: One   on 12th September
1988: One   on 20th September
1990: Two   on 19th October

These are the first records for the region.

Snow Bunting *Plectrophenax nivalis*
Regular autumn migrant and winter visitor.   This species was recorded almost annually throughout the period 1982-1991.

1982: One at Shannon on 17th October
Five at Mount Callan in late October
One at Doolin on 30th December
1983: Present from 1st October to 23rd November at Aughinish, Limerick with a maximum of eight on 29th October
1985: One at Shannon Airport Lagoon on 19th October and one in Shannon Town on 18th November
1987: Seven on 14th February at Cliffs of Moher
Six at Loop Head on 6th October with a peak of 35 in December
1988: A peak of eight at Loop Head on 9th October
1989: A peak of 10 on 30th September and 1st October at Loop Head. Recorded on nine days with the earliest record from 16th September
1990: Recorded on six days with a maximum of two on 13th and 14th October
1991: At Loop Head; three on 28th September, one on 1st October, two on 7th October, singles on 9th, 13th, 25th, 26th and 27th October
One at Quilty on 1st December
One at Aughinish, Clare on 20th January

Yellowhammer *Emberiza citrinella*
Locally common breeder in certain limestone areas such as the Burren in Clare and Barrigore area of west Limerick, but absent from large areas of east and south Limerick.   The only record received was of six birds at Limerick City on 8th January 1983.

Ortolan Bunting* *Emberiza hortulana*
Rare autumn migrant.
1988: One at Kilbaha on 18th September (P.Brennan)

This is the first record for the region of this species.

Rustic Bunting* *Emberiza rustica*
Rare northern European vagrant.
1985: One was trapped at Kilbaha on 13th October (P.Brennan)

This is only the second Irish record, and the first for the region, of this species.

Little Bunting* *Emberiza pusilla*

Rare northern European vagrant.

1988: A juvenile at Kilbaha on 16th October (P.Brennan et al)
1990: Two at Kilbaha on 13th October (T.Mee et al)

These are the second and third records for the region of this species.

Reed Bunting *Emberiza schoeniclus*

Counts of up to 50 are regular at Loop Head, and the highest count is 60. The only other record received was 41 birds at Fenloe on 8th April 1989.

Black-headed Bunting* *Emberiza melanocephala*

Rare southern European vagrant.

1991: A male at Loop Head on 17th and 18th July (J. Murphy et al)

This is only the fourth Irish record, and the first for the region, of this species.

Feral Species

The following species were also recorded during the period of the report. They are treated separately, however, as they are definately descendants of birds bred in captivity.

Black Swan *Cygnus atratus*

Two birds were seen at Lough Atedaun on 18th October 1986. At least one remained until 7th December. These birds almost certainly originated from a wildfowl collection.

Greylag/Chinese Goose *Anser anser* x *A. cygnoides*

A flock of *circa* 30 birds is now resident at Lough Gur. This hybrid flock was originally introduced by a local gun club.

Appendix

While the following species were not recorded in Clare or Limerick during the ten years of this report, they are included here in order to create a combined 'all-time' list for future reference.

Black-necked Grebe *Podiceps nigrocollis*
Two Limerick records in the present century prior to 1954.

Little Bittern *Ixobrychus minutus*
One shot at Limerick, January 1942.

Night Heron *Nycticorax nycticorax*
One found dead at Bruree, November 1957.

Glossy Ibis *Plegadis falcinellus*
Four records for Clare between 1857 and 1906.

Spoonbill *Platalea leucorodia*
Single birds at Shannon Airport Lagoon, one from 26th June 1975 to 1st July 1975, and an adult on 29th and 30th August 1976. There are five previous records.

Pink-footed Goose *Anser brachyrhynchus*
One shot in Limerick, 1930.

Ruddy Shelduck *Tadorna ferruginea*
A flock of six (of which one female was shot) was seen on the south shore of the Shannon Estuary, 1882.

Blue-winged Teal *Anas discors*
One shot on the Shannon Estuary in north Limerick in October 1980. There are two previous records.

Smew *Mergus albellus*
A female was shot at Curraghchase in February 1942.

White-tailed Eagle *Haliaeetus albicilla*
An eyrie still existed at the Cliffs of Moher in 1849, but had disappeared by 1900.

Golden Eagle *Aquila chrysaetos*
Bred in Clare up to 1896.

Little Bustard *Tetrax tetrax*
One was shot near Ennis in December 1916.

Avocet *Recurvirostra avosetta*
One at the Shannon Airport Lagoon on 23rd and 24th December 1981.

Stone Curlew *Burhinus oedicnemus*
One was shot in Clare in the autumn of 1844.

Pranticole sp. *Glareola sp.*
One was seen near Crusheen in August 1953.

Laughing Gull *Larus atricilla*
One at Shannon Airport Lagoon from 26th June to 21st August 1981.

Pallas's Sandgrouse *Syrrhaptes paradoxus*
Several near Carrigaholt in May 1888, and for some months afterwards.

Snowy Owl *Nyctea scandiaca*
An immature was shot near Ardagh in March 1907.

Roller *Coracias garrulus*
One was shot near Riverstown around 1855.

Hoopoe *Upupa epops*
One at Ardnacrusha in September and October 1957. One was shot at Quilty in 1960.

Wryneck *Jynx torquilla*
One seen at Loop Head in October 1931.

Savi's Warbler *Locustella luscinioides*
One at Shannon Airport Lagoon from 17th to 23rd June 1980.

Dusky Warbler *Phylloscopus fuscatus*
An injured bird was found near Limerick City on 5th December 1970, but died within a couple of days. The bird had been ringed at the Calf, Isle of Man on 14th May 1970.

Golden Oriole *Oriolus oriolus*
A small flock was seen near Ennis in 1847. Other records are of two at Corofin prior to 1862 and an adult female found dead at Colemanswell, Limerick in May 1947.

Great Grey Shrike *Lanius excubitor*
One was seen in the Burren in February 1971.

Dark-eyed Junco *Junco hyemalis*
One was shot at Loop Head in May 1905.

Corn Bunting *Miliaria calandra*
Formerly resident. The most recent record is of one at Kilfenora in May 1971.

The west's awake!

Rustic Bunting *(Phil Brennan)*

Phil Brennan
The Crag, Stonehall, Newmarket-on-Fergus, Co. Clare.

The morning of the 12th of October 1985 dawned grey, with a light south-easterly wind. Having slept overnight in my car, I awoke to the task of opening or erecting nets with that queasy early morning feeling. Opening the nets in Gibson's grove and orchard came first; then I started to put nets up at the sallows, which had not been possible the previous evening due to failing light.

Preoccupied at my tasks, I saw little of note. Given that the previous week had seen Loop's first rare land-bird for fifty four years, a Red-breasted Flycatcher, I still hoped that it and the two accompanying Pied Flycatchers, the second county record, wasn't just a flash in the pan. Indeed, after such a dismally wet summer when most of my efforts at ringing Sedge Warblers at Shannon had been fruitless, my very reason for being in Kilbaha was all based on hope, hope that the site just might attract migrants, hope that the Wryneck of 1931 and the American Junco of 1905 were not just history but pointed towards a more modern equivalent.

They were! After erecting the second of two sixty foot nets among the sallows, I went back to check them. Clare's first Yellow-browed Warbler and a lovely male Redstart lay in the panels. Astonished, I rushed to check all the nets and had a busy day with two Pied Flycatchers, a Garden Warbler, and an influx of Coal Tits (nineteen handled in all) to keep me busy. Between the 12th and 13th of October, fifty Coal Tits were ringed, an astounding number for such a westerly and exposed area.

The south to south-easterly winds continued gently on the 13th but the morning was quiet after the excitement of the 12th. At about 13.00hr I found the net in the orchard again full with Coal Tits. I went back to the car to get more bird bags and then extracted the birds. There was a Dunnock and a bunting also in the net. I had been working so quickly that it wasn't until I took the bunting out of the bag that I realised it wasn't the expected Reed Bunting. I ringed the rest of the birds and then looked at it again. What the hell was it? I took a good description, a sketch and some photos. By the time of release I was fairly sure I had a Rustic Bunting and subsequent references to literature proved this to be only Ireland's second record.

I saw a Yellow-browed Warbler later in the orchard, but couldn't see if it was the bird I had ringed the day before. In fact, with one hundred and fifty birds ringed in two days, I hadn't much time for watching - what would an extra observer or two have also picked up, I wonder? To me, this weekend will always remain a 'classic' Loop event, with south-easterly winds the cause of the falls. But luck must have been with me; how was it that I just happened to do my first full ringing session at Loop on the very days of Britain and Ireland's largest Yellow-browed Warbler arrival ever?

It should be remembered that given Co. Clare's westerly position, the Redstart, Garden Warbler and the Pied Flycatchers were almost as rare, relatively, as the Rustic Bunting.

The 1985 autumn produced some other good days, with Tree Pipit, Barred Warbler, three more Yellow-browed Warbler, another Red-breasted Flycatcher and a Lapland Bunting.

Land-bird watching at Loop Head is of course somewhat overshadowed by the magnificent sea-watching and manning the place is no picnic for the easily bored or even for the very patient. Its position means that while the south coast enjoys lots of 'stuff', we often get very little for a lot of effort. Despite these difficulties, or maybe because of them, the rewards when they do come are, I think, better appreciated than at stations where rarities are more regular.

The land watching at Loop Head certainly has had its moments with such wonderful birds as Arctic Warbler, Yellow-rumped Warbler, Red-rumped Swallow, Gray-cheeked Thrush, Pied Wheatear, Little Bunting, as well as Hobby, Ortolan Bunting, Richard's Pipit, Scarlet Rosefinch, Buzzard, Marsh Harrier, etc. Common birds are recorded faithfully and there have been good falls of Goldcrest, large crow movements and finches, especially Chaffinch, of which over two thousand passed through in a single moring. Long may it continue.

Charleville Lagoons: A site guide

Green Sandpiper (*Phil Brennan*)

Tony Mee
Ballyorgan, Kilfinane, County Limerick.

Location
The site is situated 1km north of Charleville, just off the main Limerick to Cork road.

Description
The site consists of a series of man-made lagoons used for the treatment and containment of liquid waste from the Golden Vale dairy plant about 1km away at Charleville (Figure 1). The treated effluent having gone through an aeration treatment process (at central lagoons 9 and 10) is pumped into the various settling beds/lagoons with an eventual outflow into an adjoining stream. This stream which divides the lagoon complex also constitutes the county boundary between Cork and Limerick.

Fluctuations in water levels
The Cork lagoons (No. 1-11) are full all year round with little visible change in water levels. The water levels of the Limerick lagoons (No. 12-16) are directly related to the dairying season. Lagoon 12, the largest and the main duck pool is always full. In contrast, lagoons 13-16 which are empty during winter and spring begin to fill during June and are usually full by late July, thus reducing dramatically their appeal to autumn passage waders. These lagoons then begin to empty out during October and are normally drained by mid December. Lagoon 18 is a shallow drainage pond which does not vary much in water level.

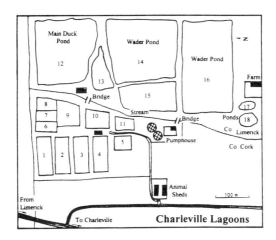

Charleville Lagoons

Vegetation
The dominant vegetation along the banks of the county Cork pools is stinging nettle *Urtica dioca* which provides good ground cover for breeding duck. The steep banks of the county Limerick pools are, in contrast, quite stony and bare.

Status
The lagoons themselves are subject to a Golden Vale 'No Shooting' order, though winter-time shooting does occur in the immediate vicinity of the lagoons causing unfortunate disturbance to the lagoons' wintering bird population.

Bird utilisation
The county Cork lagoons are the least important for birds. These lagoons normally hold the bulk of the site's Mallard population, except in late summer when a moult flock congregates on the bigger Limerick lagoons.

The Limerick lagoons attract the bulk of the site's duck and wader species. Fifteen duck species have been recorded to date at the lagoons, including such rarities as Garganey, Ring-necked Duck, Ruddy Duck and Ferruginous Duck.

Twenty-one species of wader have been noted at the lagoons, an impressive total for a site so far inland. Wader rarities here have included Broad-billed Sandpiper and Grey Phalarope. Other unusual species which have found their way to the lagoons include Leach's Petrel, Black Tern, White-winged Black Tern and Sabine's Gull. To avoid duplication, summary details as to the incidence of both the common and rarer species recorded at the lagoons can be found by reference to the ten year species list elsewhere in this report.

This article concentrates on a number of species which, for a variety of reasons, merit greater attention.

Shelduck *Tadorna tadorna*

Birds have been recorded annually in spring from as early as 23rd February. Numbers vary annually from one to four pairs. During this period, empty lagoon beds are adopted by individual pairs and a strong pair bond is always evident with frequent territorial flights, particularly when another pair or pairs are present in neighbouring lagoons. Normally birds have vacated the lagoons by early June. The exceptions being in 1978 and 1988 when individual pairs remained to breed. In 1988 a pair produced a brood of eight chicks, of which seven fledged successfully. The infrequent nature of successful breeding at the lagoons is probably due to insufficient levels of available feeding, with the presences of ground predators such as fox and mink also acting as a likely deterrent.

In Ireland there are established inland breeding populations at Lough Neagh and Poulaphouca Reservoir, Co. Wicklow (Hutchinson 1989). Charleville Lagoons situated at around 50km from the nearest coastal population is an example of this species willingness to colonise any inland site which resembles their preferred maritime habitat of tidal saltmarsh or coastal dune.

Whimbrel *Numenius phaeopus*

Recorded infrequently in small numbers on autumn passage, examples of typical records being two on 22nd July 1986 and four on 12th September 1986. Spring passage was first noted in 1989 when a flock of 40 birds was observed arriving to roost on an empty lagoon bed on the evening of 20th April.

In 1990, closer scrutiny confirmed a regular pattern of arrivals in the hours prior to dusk with most birds approaching from a low trajectory. Birds quickly settled down to roost on what were presumably traditonally favoured spring passage roost locations along the empty lagoon bed. 1990 counts were as follows: seven on 22nd April; 82 on 26th April; 248 on 4th May; 100 on 11th May; and five on 20th May.

In 1991, spring passage coincided with major lagoon reconstruction work. The birds loyalty to the site was tested by significant changes to the lagoons themselves and the overnight presence of heavy machinery, and yet, despite this, the birds continued to use the lagoons, as in previous years, in the following numbers: 32 on 25th April; 140 on 27th April; 200 on 3rd May; and 90 on 12th May.

While the pattern of spring Whimbrel passage in Ireland is well known, the occurrence of regular spring night-time roosts has not been previously documented. A specific Whimbrel ringing programme was commenced in 1990 and proved highly successful with 109 birds ringed which accounted for 81% of the entire British and Irish 1990 ringing totals for this species. A number of birds have been retrapped at the lagoons in successive springs, confirming that at least some birds are part of a return passage each spring. In addition, the ringing programme has been rewarded with a French recovery (see ringing report for details).

Green Sandpiper *Tringa ochropus*
This species has been recorded in every month with a monthly breakdown as follows:

	JAN	FEB	MAR	APR	MAY	JUNE	JULY	AUG	SEPT	OCT	NOV	DEC	TOTAL
1985	-	-	1	1	-	-	9	5	6	-	1	2	12
1986	2	-	2	-	1	3	6	8	3	2	5	3	16
1987	4	6	1	-	-	-	1	6	6	3	6	5	15
1990	4	6	4	7	-	-	6	3	-	1	3	3	15
1991	3	3	3	3	2	-	5	8	8	1	2	2	14

There is a notable autumn passage during July and August, with a percentage remaining on to overwinter. Numbers in winter generally peak in December when the back lagoons are almost empty with lots of fresh exposed mud. Numbers decline in January as the lagoon floors dry out and ground temperatures normally drop. Numbers can fluctuate greatly from year to year with 1988 and 1989 being exceptional with peaks of 14 on 15th October 1988 and 16 on 24th December 1989.

Latest spring record was two on 5th May 1991 and earliest return sighting was one on 22nd June 1986.

The lagoons are one of the principal haunts of this species in Ireland and confirm this species preference for inland wetlands.

Acknowledgements

I am deeply indebted to Golden Vale Co-operative pc. for their help and co-operation with bird studies at the lagoons.

Pied Wheatear at Loop Head - a species new to the west of Ireland

Pied Wheatear *(Phil Brennan)*

Phil Brennan
The Crag, Stonehall, Newmarket-on-Fergus, Co. Clare

Tom Tarpey
18 Greenview Drive, The Fairways, Monaleen, Limerick.

Tony Mee
Ballyorgan, Kilfinane, Co. Limerick

Introduction

At about 14.00hr on 5th November 1988, Tony Mee and I (P.B.) were driving slowly eastwards through Kilbaha, in the process of checking the mist-nets in the area. The day was cloudy, but mild, with a force two or three south-easterly wind. Outside one of the farms, just east of the orchard, despite a fogged-up windscreen, Tony noticed a 'Wheatear' standing on a pile of gravel in the 'street' of the farm. It would have been unusual to find a Common Wheatear in such a place, and the last for the season had been seen about ten days beforehand.

We got out of the car to inspect the bird. It flew to the top of the roof of the farmhouse and began flycatching. It quickly became apparent that we were dealing with an unfamiliar Wheatear species, especially when a black/grey underwing, a striking tail-pattern and unusual head-pattern and breast colour were noted. Sketches, photos. and notes were taken and after about 20 minutes we were joined by Tom Tarpey and Jacinta Reynolds. The bird was watched until dusk. Consultation with the available literature, including *Birds of the W. Palearctic. Vol. 5, Irish Birds 1982*, pp 189-192, *Irish Birds 1981*, p.113(Photo.) and various editions of *British Birds* (photos.), led to the conclusion that the bird was a male Pied

Wheatear *Oenanthe pleschanka*. When compared with the first Irish record of this species at Knockadoon, Co. Cork in November 1980, the only apparent difference was that the Kilbaha bird did not have as bright a supercilium.

Description

Structure, Size and Jizz The bird was a little smaller than Wheatear *Oenanthe oenanthe*, and though structurally similar, was quite chat-like. Its shorter legs also contributed to this impression. The bird spent much time flycatching from the roofs and gables of the farmhouses and outbuildings. It was very active most of the time. In a most un-Wheatear like fashion, it even flew into some of the sheds on the second day and probably roosted there. In flight, its tail and underwing patterns were the most striking feature. It didn't seem very shy, and even flew over the watchers while flycatching.

Head Crown and nape brown (tone difficult to describe, perhaps slightly bronze). This brown merged into a subtle supercilium. The brown on the crown almost, but not quite, reached the upper mandible. The supercilium was indistinct at a distance but closer up it was more obvious, coloured pale creamy-brown, reaching fairly well back, and this colour seemed to cover the area just to the rear of the ear-coverts. Its outline, however, wasn't very clear-cut at any range. Lores and eye-stripe were black, but the stripe only extended to above the ear-coverts, narrowing as it did.

The ear-coverts had a blackish base, this covered with creamy-brown and whitish fringes. This pattern gradually became more black towards the front of the face (under the eyes and lores) and throat, but still with some pale fringing. This darkish area formed a clear border with the upper breast and sides of neck, and there was a very narrow whitish band separating these two zones.

An imperceptible narrow pale eyering was apparent on close inspection.

Back Nape and mantle the same colour as the crown, with pale fringing that was only obvious at very close quarters. This colour ended abruptly high on the upper rump. The upper rump was greyish-white but most of the rump was white as were the upper tail-coverts.

Underparts Upper breast orange, similar to Stonechat, fading to much paler whitish with a buffy-orange suffusion on the belly and flanks, with the vent palest.

Tail White overall, but with a terminal black band. This band was fairly narrow but the tips of the outer feathers had slightly more black, and the outer feather had even more, forming a slight border of black on the outside of the feather from a quarter to one-third way up the feather. The central tailfeathers were completely black. At rest, this gave a predominantly black impression to the tail, but when the bird flew, the more extensive white area was revealed. The undertail was generally whitish, with the black terminal bar showing.

Wings The upperwing was patterned more-or-less as Wheatear, with the feathers black-based. All the feathers had varying amounts of fringing. This fringing was buff-brown. The fringes were broadest on the tertials, the greater and median coverts. However, they were generally narrower and slightly less obvious than in common Wheatear. The alulal area was blackest and the least fringed.

The underwing was very distinctive. It gave a blackish appearance overall, but close views showed the underwing coverts to be completely black, whereas the primaries and secondaries were of a grey-black tone.

Bare parts
Eye - seemed blackish.
Bill - black.
Legs - black, shorter than Common Wheatear.

Confirmation
The main features which confirmed this bird as being a Pied Wheatear were:
(1) Head pattern and colour, especially the supercilium and the facial pattern;
(2) The two-toned underwing colouration;
(3) Colouration of breast and underparts;
(4) Pattern of black-and-white on tail and rump;
(5) Length of legs;
(6) Size.

Birds of farmland in Clare and Limerick

Liam Lysaght
Poll na Lobhair, Kilnaboy, Co. Clare

Introduction

One seldom considers the importance of farmland for birds, yet farmland comprises over 80% of the total area of Limerick and Clare, making it by far the most extensive habitat available to breeding birds. The total area of farmland in the two counties is *circa* 612 000ha, and a recent study of the breeding birds of farmland in the region found that on average every 10 hectares supported 27 breeding pairs of birds (Lysaght 1989). This means that farmland in Clare and Limerick supports well in excess of one million breeding pairs.

Typical famland birds

Twenty two species are widespread and could be regarded as typical of the farmland bird community of the region. These are listed below:

Wren	Song Thrush	Bullfinch
Robin	Starling	Woodpigeon
Blackbird	Jackdaw	Swallow
Dunnock	Great Tit	Spotted Flycatcher
Chaffinch	Greenfinch	Meadow Pipit
Willow Warbler	Coal Tit	Reed Bunting
Blue Tit	Mistle Thrush	
Goldcrest	Long-tailed Tit	

Of these, 5 species are dominant, Wren, Robin, Blackbird, Dunnock and Chaffinch which together account for between 60 to 70% of the total population. In most years Wrens are by far the most abundant species comprising a fifth of all breeding birds. Some years, however, Wren populations are lower because, being so small, this species is very susceptible to cold weather. Following severe winters, some other species will be commonest, perhaps even the Willow Warbler. The four other dominant species represent around 10% each of the total breeding bird population.

In addition to the five dominant species, Willow Warbler, Blue Tit, Goldcrest and Song Thrush are also extremely widespread, though they seldom occur in great densities. At the other end of the spectrum there are a host of species which can be found breeding at low densities in farmland. Species such as the Treecreeper, Goldfinch, Linnet and Whitethroat are never very common but they are probably evenly distributed throughout the region. Other species such as Chiffchaff, Skylark, Grey Wagtail, Moorhen and Snipe also occur at low densities on farmland, but these species have fairly specific breeding requirements that are seldom satisfied on farmland, but where the habitat is suitable they are likely to be present.

The farmland habitat

In the strict sense farmland is not a habitat but rather a mosaic of different habitats; fields, hedgerows, farm buildings, streams, scrub, etc. all combining to provide a rich and varied environment for birds. The single most important habitat is the many miles of hedgerows that enclose the fields. Over 90% of all the birds that breed in farmland are dependent on hedgerows as nest sites and food sources. Not all hedgerows are equally important for songbirds, generally the best hedgerows are those with some tall trees, a variety of different shrub species and a variety of flowers growing at the base. Hedgerows that are kept neat and tidy and trimmed into a box shape are avoided by almost all species, except Dunnock, which appear to be the exception.

The next most important farmland habitat, and one which is very often overlooked, is the farmyard and buildings. On average about 4% of all farmland species breed in, or around, the farmyard. Perhaps the Swallow is the species which most comes to mind, as it nests in the rafters of many farm buildings. Other species are also strongly associated with farmyards, for example, Barn Owl, Starling, Pied Wagtail, Jackdaw, Mistle Thrush and Collared Dove. The Barn Owl needs special mention as it is a species fortunately which finds suitable breeding and feeding habitat in many of the older more traditional farmyards commonly found throughout the region. This is in marked contrast to many parts of Britain, for example, where the renovation of old farm buildings has led to a dramatic population decline in this species.

The open fields are also a separate habitat which is of attraction to a few species, such as Meadow Pipit, Skylark and Pheasant. In this region, where grassland is the dominant, and almost exclusive use of the land, field species are poorly represented, comprising only about 1% of the total bird population. Indeed, the Meadow Pipit is the only widespread field species here, although in other parts of Ireland, particularly where some land is tilled, field species would be more common.

The final main habitat within farmland is wetland, either streams, pools or waterlogged patches. The most common wetlands in farmland are streams and drainage ditches, and where these occur Grey Wagtail, Moorhen and Reed Bunting are likely to be found. In any small pond or pool, Mallard and Little Grebe may breed, and in more extensive marshy areas, species such as Snipe and Sedge Warbler may be found. Generally the larger the wetland, and the closer it is to other wetland areas, the more varied the bird life is likely to be.

In addition to the above, roads, bungalows, stone walls, railways, graveyards and many other features also contribute to the rural landscape and should be regarded as habitats providing opportunities for birds to exploit. Thus, in addition to being by far the most extensive habitat in the region, farmland is a very rich and varied environment for our birds, and more attention should be given to its conservation.

Seawatching at the Bridges of Ross

Great Skuas at the Bridges of Ross *(Phil Brennan)*

Tony Mee
Ballyorgan, Kilfinane, County Limerick.

Background

Affectionately known as 'the Bridges' by birdwatchers throughout the country, the Bridges of Ross on the Loop Head peninsula in west Clare is now firmly established as the premier seawatching point along Ireland's west coast - but this was not always the case. One does not have to go too far back in time in order to trace the beginnings of seawatching at 'the Bridges'.

In the late seventies, Philip Buckley and Ewart Jones commenced seawatching at Loop Head itself. From 1979 to 1984, their dedicated autumn coverage gradually switched from the Head itself to the nearby Bridges of Ross. This site is lower to the sea permitting much better views as the birds follow the contours of the Clare coast before rounding the Head. With a resurgence of interest nationally in seawatching, the mid to late eighties saw an increase in the number of seawatchers visiting the Bridges from throughout Ireland and the United Kingdom. Regular sightings during this period of species such as Leach's Petrel and Sabine's Gull and later of Long-tailed Skua (an extreme rarity in Ireland) firmly established the Bridges as a major seawatch point. But enough of the 'sales- spiel', where exactly is the seawatch point?

Location

The point where the majority of seawatching is done is indicated in Figure 1. For those who are not familiar with the area, the point can be reached by taking the path to the natural bridge from the carpark and by veering

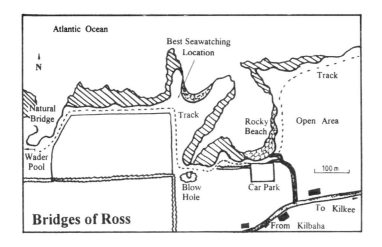

Atlantic Ocean

Best Seawatching Location

N

Track

Natural Bridge

Track

Rocky Beach

Open Area

Wader Pool

100 m

Blow Hole

Car Park

To Kilkee

Bridges of Ross

From Kilbaha

right off this path at the point where the path swings to the left at the corner of an adjoining field. Here a shallow semi-circle or mini-amphitheatre formed by weather erosion close to the machair edge gives the watcher a modicum of windbreak and a wide field of view, particularly to the north (right) which is the direction most birds will approach from during autumn seawatches.

Weather conditions

When strong onshore winds occur during autumn, then the chances are increased of witnessing a brisk seabird passage. The Bridges are no exception. On north-west or west winds, seabirds moving south and tracking into the wind can get pushed in closer to shore than normal, giving the watchers an opportunity to gain close-up views of a wide cross-section of seabirds.

The best seawatches have taken place during north-west gales particularly if the storm centre lies well to the north over, or close to northern Scotland. A deepening low over this area causes a dispersal of birds from these northern waters south, to be funnelled down along the western seaboard. To illustrate this point, a weather chart for the 2nd September 1988 is shown (Figure 2).

On this date during force 6 to 7 westerlys some of the birds seen at the Bridges were four juvenile Long-tailed Skua, six Sabine's Gull (four juveniles and two adults), 64 Leach's Petrel, and a supporting cast of 23 Arctic Skua, 20 Great Skua, 900 Sooty Shearwater, 500 Storm Petrel and five phalarope spp.

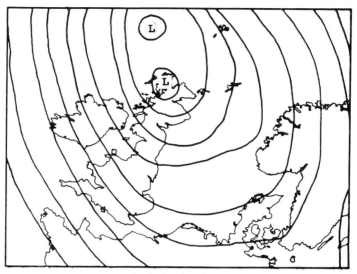

2nd September 1988, 07.00 hrs.

Seasonal variation

July and August have been relatively under-watched when compared to September and October. August provides the best prospects of seeing a Great or Cory's Shearwater and this month also claims both site records for that pelagic rarity, Wilson's Petrel.

West coast seawatching produces a distinctive avifauna when compared to the south and east coasts. This is best illustrated in September and October when a significant skua passage occurs, Sabine's Gull are regularly recorded and Leach's Petrel often outnumber Storm Petrel. A day total of 337 Little Auk in December 1991 showed the potential for recording this species during the winter months. Spring passage seems to be confined in the main to our common breeding seabird species but very little intensive seawatching has been done at this time of the year.

Wilson's Petrel off the Bridges of Ross

Killian Mullarney
Redshire House, Murrintown, Co. Wexford

Introduction

It was not until we heard the radio weather forecast on the morning of 5th August 1990 that Oran O'Sullivan and I realised there was a likelihood of good seawatching conditions at Bridges of Ross that day. A deepening low pressure system of 988 HPA was passing in a northeasterly direction over the northern half of Ireland with a strong west to northwest airflow in its wake.

After a four hour drive from Wexford we arrived at 13.30 GMT to find the wind blowing force 5 to 6 from the west, and an excellent passage of common seabirds in progress. Manx Shearwaters were streaming past (at a rate of up to 2000 per hour) with Fulmars and Kittiwakes being the next most numerous species (up to 800 and 500 per hour respectively). Not so obvious, but similarly numerous, were Storm Petrels which systematic counts revealed were passing at a rate of up to 500 per hour. It was on these that I focussed most of my attention.

Most Storm Petrels were passing within about 400m and with the excellent light conditions that afternoon, it was possible to discern the white underwing bar on even the more distant birds. An hour and a half into the seawatch I picked up a petrel that immediately registered as being 'different'; the bird was close, about 150m away, and apart from an impression of it being larger than a Storm Petrel (though none was directly alongside for comparison) I registered a plain-looking underwing and a conspicious pale upperwing band. I immediately thought it might be a Leach's Petrel and I alerted O. O'S. who was right beside me. I continued watching the bird through a 20x telescope as I called out directions to Oran. After a few more seconds I realised the bird did not look right for a Leach's Petrel either, as it did not have the 'arm-length' of Leach's. Subconsciously, I suppose, I had put this down to a temporary effect of foreshortening but, as the angle of view changed, I gained a better impression of the wing shape. Suddenly it was clear; this was not a Leach's Petrel, it was a Wilson's! Fortunately, Oran also located the bird but it quickly became obscured from sight by a rapidly approaching heavy squall. I had observed the bird for about 25 seconds, and the account that follows is based on notes and sketches made immediately after the sighting.

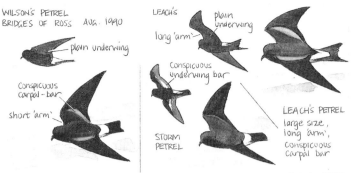

Handwritten annotations on illustration:

WILSON'S PETREL
BRIDGES OF ROSS AUG. 1990

plain underwing

Conspicuous carpal-bar

short 'arm'

LEACH'S
plain underwing
long 'arm'

Conspicuous underwing bar

STORM PETREL

LEACH'S PETREL
large size,
long 'arm',
conspicuous carpal bar

Petrels (*Killian Mullarney*)

Description

Shape and size The bird was similar to a Storm Petrel but it immediately registered as different. It seemed more robust and perhaps a little larger than a Storm Petrel.

Plumage During this seawatch, we saw hundreds of Storm Petrels, which all appeared to have plain upperwings, lacking the thin whitish wing-bar often shown by juveniles later in the autumn. Our bird showed striking plumage differences with a conspicious greyish carpal-bar and a complete lack of Storm Petrel's distinctive white underwing bar. The upper and underwing patterns were therefore more reminiscent of Leach's Petrel.

Wing shape The wing shape was distinctly different from Leach's. Most of the time the wings appeared scythe shaped, broad at the base, pointed at the tip and strongly 'swept-back', rather *Hirundine*-like in fact. The leading edge of the wing seemed strongly curved but lacked the distinct angle at the carpal shown by both Storm and Leach's Petrel.

Tail and rump The tail appeared to have a square or slightly rounded end, much like a Storm Petrel. The shape of the white rump was very like a Storm Petrel's, with the white extending prominently onto lateral undertail coverts. It seemed to show a more conspicuous 'rump' than the Storm Petrels, something which has not struck me particularly during previous encounters with Wilson's Petrels off boats. I looked for, but could not see any sign of toes projecting beyond the tail.

Flight and behaviour The bird's flight-action, or rather the lack of it, was remarkable. Throughout the 60m or so it travelled while in view, heading more or less directly into the force 5 to 6 westerly wind, the bird did not seem to flap. I am used to seeing both Leach's and Storm Petrels 'hang' on stiff, slightly raised wings, but the sight of this bird travelling so effortlessly was amazing. It glided more or less level, just above the waves, on still, almost horizontal wings. It must have banked from time to time

83

since its plain underwing was apparent. Just before it disappeared from view, it performed a series of extended 'hops', seeming to kick the surface of the water with its feet (which I could not actually see). It hopped three or four times along a straight course, covering a foot or two between each hop.

Discussion

The combination of plain underwing, conspicuous pale greyish carpal-bar, wing shape, extensive white 'rump' and the remarkable flight action allowed a confident identification of the bird as a Wilson's Petrel. In retrospect, it was perhaps the flight-action that was most eye-catching since this differed markedly from the much 'busier' flight of the many Storm Petrels passing at the same time. Since 1990, I have watched Wilson's Petrels passing close to shore off North Carolina, U.S.A. and at sea between Portugal and Madeira. In both cases I observed birds progressing in the same distinctive way as noticed at the Bridges of Ross. Wilson's Petrel is not too difficult to distinguish from Storm and Leach's Petrel provided a good view of the bird is obtained. As both of the latter resemble Wilson's in different ways, thorough familiarity with them is essential.

Pelagic trips off south and west Ireland, and southwest Britain since the mid 1980s have proven that Wilson's Petrels are out there, albeit in small numbers, in the early autumn. They generally associate with Storm Petrels, and it may well be that the occasional late summer/early autumn storm that pushes large numbers of Storm Petrels into inshore waters provides a real chance of seeing Wilson's Petrel *from land* in Ireland. It could be that by September, when storms are more frequent and more time is put into seawatching the bulk of Wilson's Petrels that pass through Irish waters are too far south to be affected by our weather systems.

This is the second record of Wilson's Petrel at the Bridges of Ross. The first was seen by Philip Buckley on 18th August 1985 (O'Sullivan and Smiddy 1988). More than just a coincidence?

A ringing study of Black-headed Gulls breeding on Lough Derg

Black-headed Gull (Phil Brennan)

Tom Tarpey
18 Greenview Drive, The Fairways, Monaleen, Limerick.

Introduction

The Black-headed Gull is a widespread and abundant species throughout northern and central regions of Europe and Asia. Its breeding range extends westwards to Iceland and eastwards as far as the Kamchatka Peninsula in Siberia, and typically between the latitudes of 43 ^{0}N and 66 ^{0}N. It is partly resident, and partly dispersive and migratory throughout its range. In general the non-breeding range extends southwards from the breeding areas into Mediterranean regions, north Africa and southern Asia (Harrison 1982).

In Ireland the breeding range exhibits a similar northern bias with the bulk of colonies generally located north of a line joining the Shannon Estuary and Wicklow Head, with the greatest concentrations in the mid west, west and north east (Gibbons et al 1993). The winter range shows a more widespread distribution with greater numbers generally near the east and south coasts (Lack 1986). The wintering population is known to be boosted by a substantial immigration of British and Continental birds as well as by recruitment from the previous breeding season. Post breeding dispersal is generally considered to be confined within the island with relatively small numbers of emigrants (Hutchinson 1989).

Lough Derg is located within the main breeding zone for Black-headed Gulls in Ireland. In the early eighties it held up to 31 separate colonies ranging in size from a single pair to four hundred pairs with a maximum total count of almost 2200 nests (Reynolds 1990). This paper presents the results of a ringing study on part of the Lough Derg population over a seven year period from 1983 to 1989.

Study sites

The study sites are located at the two largest concentrations of breeding gulls on the lake, i.e. The Cormorant Islands, near Scarriff, Co. Clare and Rinn Island, Ryninch, Co. Tipperary (Figure 1). The Cormorant Islands consist of a group of rock outcrops. The largest island has some soil cover with tussocks of grass and nettles. The other islands are almost devoid of vegetation. The maximum nest count was estimated at 548 in 1985. Rinn Island is covered with rough grass, nettles and scrub vegetation and has small areas of exposed rock and some gravel fringes. The maximum nest count was estimated at 400 in 1985.

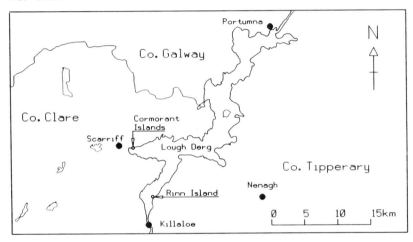

Figure 1. Location of study sites

Methods

Young birds were ringed on varying dates between May 30th and July 9th over the seven years of the study. However, the vast majority were ringed within the first two weeks of June each year. All of the birds ringed were between the post nestling and pre-fledging stages ranging in approximate ages from two weeks to five weeks old. A total of 1252 birds was ringed between the two locations. The annual breakdown of each site is shown in Table 1.

Table 1. Annual Ringing Totals at two sites in Lough Derg.

Site	1983	1984	1985	1986	1987	1988	1989	*Totals*
Rinn Island	170	182	144	128	-	-	-	624
Cormorant Islands	-	-	150	170	100	98	110	628
Combined Annual Totals	170	182	294	298	100	98	110	1,252

The fledging rate of ringed birds was presumed to be generally greater than 95%. The colonies were checked for fatalities during late June and early July in each of the first four years of the study. The percentage of ringed birds recorded dead at the colonies was generally less than 2% and never greater than 5%.

Results and discussion

Recovery rate

Fourteen recoveries have been reported up to September 1993, giving a recovery rate of 1.12%. This rate is considerably lower than the overall recovery rate for the British Isles and Ireland which is 4.6% (Mead 1974). The difference in rates may be partly explained by relatively small numbers of birdwatchers and low population density in Ireland compared with Britain. The level of awareness of ringing among the general public in Ireland may also be a factor. The rate compares more favourably with the recovery rate of 1.88% for birds ringed in Ireland between 1980 and 1991. Given that the species is relatively long lived, there are possibilities for a higher ultimate recovery rate.

There was a considerable difference between the recovery rates of the two colonies. The Cormorant Islands produced a rate of 1.43% compared with a rate of 0.80% for Rinn Island with almost identical numbers of birds ringed at both sites.

An analysis of the timing of recoveries within the calender year is given in Table 2. The majority (67%) were recorded in the period between July and September with the remainder in November and December. There was a complete absence of recoveries for adult birds in the first half of the year. Given that the December recoveries were both sight records, it is somewhat surprising to find most of the remaining recoveries clustered together in the late summer/early autumn period rather than in winter when birds might be expected to come under pressure from severe weather. However as the total number of recoveries is relatively small it may not be valid to draw any inference from this observation.

Table 2 Timing of recoveries of Black headed Gulls ringed at natal colonies in Lough Derg

	Jan.-Jun.	July	Aug.	Sept.	Oct.	Nov.	Dec.	Total
Juvenile Birds (< 3 Months)	1	2	2	-	-	-	-	5
Adult Birds (> 1 Year)	0	2	2	2	-	1	2	9
	1	4	4	2	0	1	2	14

Mortality and cause of death
The majority of recovered birds (64%) had survived at least into their second year. The recovery duration varied betwen 0.04 years (21 days) and 5.15 years, with an average recovery duration of 1.65 years.

Ten of the birds (71%) were recovered dead or dying, of the remainder two were sight records and two were found injured and subsequently rereleased.

Most of the recovery notices gave little or no information on the cause of death. However at least three of the ten fatalities recorded in this study were related to human activities. In these cases the cause of death were: trapped under a cattle grid; caught in a football net and a road casualty.

Movements and dispersal
An analysis of the distance moved by birds from the study area is given in Table 3. The majority of birds (86%) were recovered over 10km away from their natal colonies with half of those birds being recovered at distances of over 100km. The average distance moved was 74km. The lenghts of movements recorded in this study are very similar to results reported from the country in general. The quickest movement was one of 103km, from the Cormorant Islands to Donoghmore in Co. Cork, in 33 days.

Table 3 *Distance moved by Black headed Gulls from natal colonies in Lough Derg, Cos. Clare and Tipperary compared with Ireland in general.*

Distance moved	Lough Derg		Ireland (1980 - 1991)	
	n	%	n	%
0 - 9 km	2	14.2	32	21.5
10 - 99 km	6	42.9	62	41.6
100+ km	6	42.9	53	37.6
Total	14	100	149	100

Source   Lough Derg: This Study
            Ireland: Forsyth (Irish Ringing Reports, 1980 to 1991)

The full distribution of recoveries is shown on Figure 2. The dispersal pattern indicates that most birds (71%) were recovered south of the breeding sites. There were no recoveries of juvenile birds north of their natal colonies. A comparison of the *New Breeding Atlas 1988-91* and *Winter Atlas* distribution maps for Black-headed Gulls indicates a post breeding movement away from the breeding areas in the west and north of the country to southern and eastern parts. The results of this study indicate a broadly similar trend, with a particularly strong bias towards the south. The absence of any foreign recoveries to date supports Hutchinson's (1989) suggestion that there is little emigration of Irish birds.

Cramp et al (1982) indicate that British birds are highly faithful to

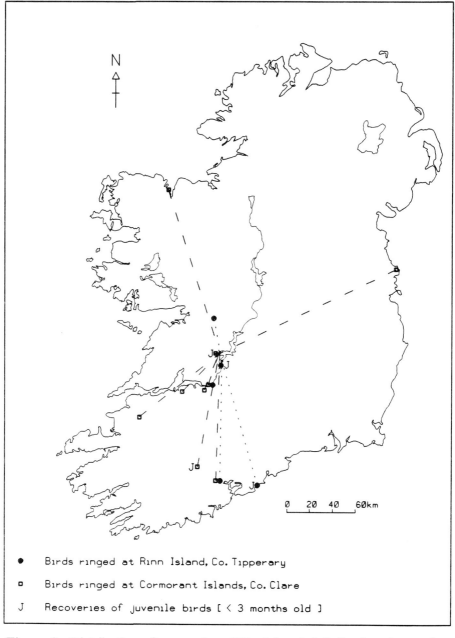

Figure 2. Distribution of recoveries of Black headed Gulls ringed at study sites on Lough Derg

their natal areas and that maximum winter dispersion is achieved in the first year by birds from southern England and Wales, and in the second year by birds from northern England and Scotland. The indication from this study is that the Lough Derg birds reach maximum dispersion in their second year, similar to their counterparts in northern Britain. Mature birds showed declining winter dispersion with age, suggesting a degree of faithfulness to their natal areas.

Birdwatching sites in Clare and Limerick

Ballyallia Lake (*Phil Brennan*)

Introduction

Throughout Clare and Limerick there are numerous areas which are good for birdwatching. We are fortunate to have an extensive coastline, the Shannon and Fergus Estuaries, many wetlands and relatively low intensity farmland over much of the region, that provide ideal habitat for birds, and consequently, for birdwatching. The experienced birdwatchers will always have their own 'personal' birding spots but for those who may be beginning birdwatching, or who may not be familiar with the region, we have described ten different locations which are well established sites. These are:

Shannon Airport Lagoon	The Burren
Ballyallia Lake	North Clare
Loop Head Peninsula	Limerick City
Lurga Point, Quilty	Lough Gur
Cliffs of Moher and Liscannor Bay	Curraghchase

The individual site guides provide a brief overview of the area, a short summary of the species associated with the site, a few tips to make the best of your birdwatching visit and, for some, a map. For brevity, only the species which are most associated with the site are mentioned, placing emphasis on the rarer or more unusual species. Nevertheless, all the sites have a wide range of common birds and are worth visiting at almost all times of the year.

Unless otherwise stated, all land is privately owned so we recommend that the permission of landowners is always sought before entering land. When birdwatching, care should be taken not to cause undue disturbance or interference to birds, particularly during the breeding season.

Shannon Airport Lagoon

Shannon Airport Lagoon is a shallow artificial lagoon, created by the building of two embankments from Dernish Island to the mainland. The lake is fringed by *Phragmites* reed beds, and bordered to the south by the extensive mudflats of the Shannon Estuary and to the north by a narrow stretch of alder and willow scrub.

Winter Whooper Swan, Pintail, Shoveler and Gadwall can occur in large numbers, and estuarine species such as Scaup and Long-tailed Duck are regular visitors. Huge flocks of Knot, Dunlin and Black-tailed Godwit can be seen on the mudflats. Rarities recorded here have included White-rumped Sandpiper, Wilson's Phalarope and Avocet.

Summer Sedge Warbler, Grasshopper Warbler and Water Rail breed in the reedbeds and adjacent alder and willow scrub. In late summer hundreds of Sedge Warbler and smaller numbers of other species use the reedbeds as a migration staging post. Rarities recorded in the past include Savi's Warbler, Bluethroat and Reed Warbler.

Birdwatching Good views of the birds both on the lagoon and on the main mudflat can be obtained from the causeway to Dernish Island. There is a display board here, describing the more common birds of the area, and so is ideal for beginners. Visit the site when there is an incoming tide as wading birds then move much closer to the causeway.

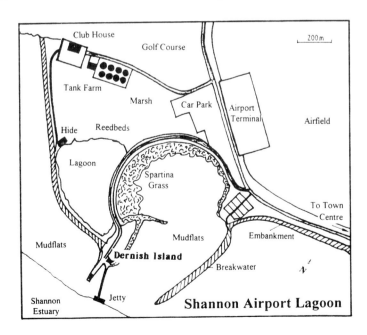

92

Ballyallia Lake

Ballyallia Lake lies on the outskirts of Ennis, just off the Galway road. It is a shallow lake surrounded by lowlying wet grassland and a small wooded area to the north. With the exception of a public carpark at the eastern end of the lake, all the land bordering the lake is privately owned.

Winter The lake is a site of national importance for Gadwall, but there are also large numbers of Shoveler, Pintail, Wigeon, Tufted Duck and Pochard. Rarer species seen in the past include Blue-winged Teal and American Wigeon. Whooper and Bewick's Swan and Black-tail Godwit frequent the flooded margins of the lake. Brambling and Blackcap have been recorded in the wood at the northern end of the lake.

Summer Great Crested and Little Grebe, Coot, Moorhen and Mallard are common breeding species around the reed fringes of the lake. Long-eared Owl have bred in the nearby woodland.

Birdwatching The main lake is best viewed from the public carpark, but there are a few spots along the small road circling the lake from which good views of swans, waders and dabbling duck can be obtained. Ideal for beginners.

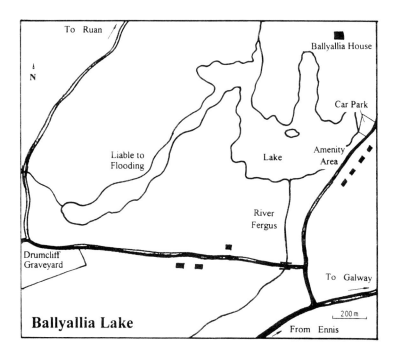

Ballyallia Lake

93

Loop Head Peninsula

Loop Head is the premier birdwatching site in the region for observing the more unusual rarities and experiencing migration at first hand. The village of Kilbaha is the focal point where the passage of migrating birds can be monitored by checking the sparsely covered gardens along the lighthouse road. Bridges of Ross, just north of Kilbaha village, is one of the best sites on the west coast of Ireland for seawatching.

Autumn September until the end of October is the main migration period, when large flocks of finches and thrushes are regularly seen passing down the peninsula. Influxes of Willow Warbler, Chiffchaff, Goldcrest and Swallow can also occur. In addition, raptors such as Peregrine, Merlin and even the occasional Hobby may be seen. The more unusual migrants include Redstart, Black Redstart, Pied Flycatcher, Garden Warbler, Snow Bunting, Lapland Bunting and Yellow-browed Warbler. Over 200 species have been recorded during the last 10 years, including rarities such as Gray-cheeked Thrush, Black-headed Bunting, Richard's Pipit, Yellow-rumped Warbler and Scarlet Rosefinch. Seawatching at the Bridges of Ross is likely to be equally rewarding, as a large number of rarities have been recorded in the past. Leach's Petrel, Long-tailed Skua and Sabine's Gull are just some of the species which occur regularly.

Birdwatching Because of the variety of species seen and the real chance of encountering the unexpected, a considerable amount of experience is necessary to make the most of the site. Nevertheless, a walk here at almost anytime of the year can be very enjoyable.

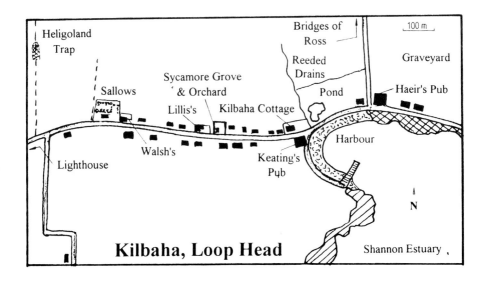

Kilbaha, Loop Head

Lurga Point, Quilty

Lurga Point is a stretch of rocky and sandy coastline lying less than 4km to the south-west of Quilty. A small road running parallel to the shore leads to the small pier at Lurga Point. From the point, Mutton Island to the north-west and the smaller Mattle Island to the south-west can be seen.

Winter Quilty is best known for its nationally important flock of up to 200 wintering Purple Sandpiper. Dunlin, Turnstone, Ringed Plover, Sanderling, Grey Plover and Oystercatcher are also present in large numbers. A variety of gull species occur, including occasionally the rarer Glaucous and Iceland Gull. Offshore, distant views can be had of the resident flock of over 300 Barnacle Geese on Mutton Island.

Summer Shag, Cormorant, and possibly a small colony of Storm Petrel breed on Mattle Island, whereas there are breeding colonies of Common, Herring, Great Black-backed and Lesser Black-backed Gulls on Mutton Island, and smaller numbers of Shelduck, Ringed Plover and Black Guillemot.

Spring & Autumn Rarer waders can be seen on passage including Common Sandpiper, Curlew Sandpiper, Little Stint and Ruff. Passerines seen in the past include Snow Bunting, Pied Flycatcher, Twite and Black Redstart.

Birdwatching From the pier at Lurga Point, exceptionally good views can be had of most of the waders. This is an ideal location for birdwatchers who wish to brush up on the intricacies of wader identification.

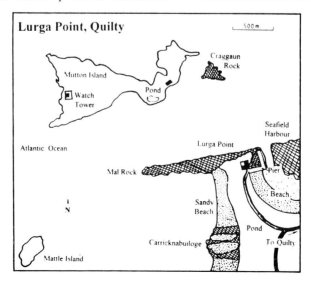

95

Cliffs of Moher and Liscannor Bay

The Cliffs of Moher, situated approximately 10km west of Ennistymon, extend for 8km and are 122m to 196m high. A carpark and visitor centre is provided close to the cliffs. Further south lies Liscannor Bay, a sweeping sandy bay stretching from the small pier at Liscannor to Lahinch. To the landward side is the Lahinch/Ennistymon marsh, a wetland which floods in winter.

Winter Great Northern, Black-throated and Red-throated Diver are regularly seen in Liscannor Bay, as are Long-tailed Duck and Common Scoter. The rarer Velvet and Surf Scoter and Slavonian Grebe have also been seen in the past. At the Lahinch/Ennistymon marsh Brent and Greenland White-fronted Geese occur occasionally, and rarities such as Black-winged Stilt have been recorded in the past.

Summer The Cliffs of Moher support Clare's largest breeding seabird colonies. Here thousands of Fulmar, Kittiwake, Razorbill and Guillemot nest on the sheer cliffs. Immediately out from the cliff at the main viewing point, lies the diminutive Goat Island where Puffins breed. Chough, Peregrine, Raven, Rock Dove and Twite also breed in the vicinity, with the latter species being an extremely rare breeding bird in Clare.

Birdwatching The Cliffs of Moher provide spectacular views of breeding seabirds in May and June, a must for birdwatchers of all levels. At other times of the year, however, very few birds can be seen there.

The Burren

The Burren is a 500km^2 limestone area in north Clare, extending from Corofin and Lisdoonvarna in the south to Black Head and Kinvara in the north. Bare limestone pavement and expanses of low hazel scrub are interspersed with turloughs. An extensive series of lakes and other wetlands lie at the south-eastern fringe of the Burren, into which most of the rainfall of the eastern Burren drains.

Winter   In the east Burren wetlands, Mute and Whooper Swan occur in internationally important concentrations. Wigeon, Lapwing, Dunlin, Black-tailed Godwit and Goldeneye are also very numerous. There is a resident flock of Greenland White-fronted Geese, and occasionally Greylag and Bewick's Swan may be seen.

Summer   Breeding Cuckoo and Whitethroat are plentiful, as are Yellowhammer, Stonechat and Wheatear. In the past Nightjar and Whinchat have bred. The turloughs support breeding Redshank, Common Sandpiper, Ringed Plover, Snipe and Lapwing. Common and Black-headed Gull breed in the east Burren wetlands, as do Hen Harrier, Shoveler and possibly Wigeon. Rarities seen in the recent past include Savi's Warbler, Garganey, Black Tern and Osprey.

Birdwatching   A walk in any part of the Burren in summer is interesting. Virtually all of the large lakes of the east Burren wetlands have carparks, from which excellent views of birds can be obtained. A visit to Lough Atedaun and Lough Inchiquin both near Corofin is recommended.

North Clare

The coastline of north Clare, stretching from Black Head in the west to Aughinish Island in the east, is generally rocky though with some exposed mudflats at low tide.

Winter This area provides the best opportunites in the region of seeing the rarer species of grebe and diver. Slavonian and Red-necked Grebe in addition to Black-throated and Red-throated Diver have all been recorded in recent years. The bay also supports good numbers of Brent Geese, Common Scoter and Long-tailed Duck. Gulls are also attracted to the area, including rare species such as Glaucous, Iceland, Ring-billed and even the extremely rare Ross's Gull.

Summer Shelduck, Red-breasted Merganser, Ringed Plover and Black Guillemot, in addition to Common, Arctic and Sandwich Tern, are just some of the species that breed along the coast. Whitethroat and Chough also breed here at present, whereas Nightjar and Whinchat have done so in the past.

Birdwatching This coastline provides excellent birdwatching opportunities throughout the year. Good views of most species can be obtained from the two piers at Ballyvaughan (old and new), at the Rine north of Ballyvaughan and from the Mortello Tower at Aughinish Island.

Limerick City

In common with most urban areas, Limerick City is not normally associated with birds, yet there are a number of interesting sites within the confines of the city. The most important feature is the River Shannon which provides feeding opportunities and a migration corridor for birds. Westfields marsh, transversed by the new link road to Clare is the best known site for birdwatching, but St. Thomas's Island, Corbally also support a good variety of birds.

Winter Teal, Tufted Duck, Shoveler and Pochard are plentiful at Westfields, and rarities including Leach's Petrel, Ring-billed Gull and Green-winged Teal have been recorded here. A variety of gulls, including rarer species such as Ring-billed, Little, Iceland and Glaucous have been recorded at the Salmon Weir above Thomond Bridge. A variety of wildfowl can be seen at Thomas's Island, but the area is best known for the large number of Cormorant which reside there. Black Redstart and Blackcap are regularly recorded in winter in the city area, whereas rarities such as Dusky Warbler and Waxwing have been seen in the past.

Summer Most of the common wetland species breed at Westfields, including Great Crested Grebe, Tufted Duck, Water Rail and Sedge Warbler. There is a colony of Grey Heron along the North Circular Road, and Swift are plentiful in the city. Rarities such as Black Tern have been seen in early summer.

Birdwatching Westfields is by far the best place to watch wetland birds as a wide variety of species can be easily observed at close range. There is also a display board showing the common species of the marsh. This site has enormous educational potential.

Lough Gur

Lough Gur is the largest freshwater lake in Limerick, and is by far the most important for wildfowl. The area is renowned as an archaelogical site, consequently the eastern end now has a visitor centre and carpark.

Winter Maximum numbers of Wigeon occur here in January and February with up to 1500 birds present. It also supports good numbers of Tufted Duck, Coot and Teal, with smaller numbers of Shoveler, Pochard and Gadwall. Whooper Swan feed in the surrounding area, but move into the lake to roost each evening. Rarer species seen here in the past include Ring-necked Duck and Ferruginous Duck.

Summer The majority of the area's breeding birds reside in the secluded marsh at the southern end. Mallard, Teal and Tufted Duck breed here in addition to Sedge Warbler and Water Rail. In the past Blackcap and Long-eared Owl have bred in the mature woodland by the carpark.

Birdwatching Good views of most of the birds can be obtained from near the graveyard at the southern end and from along the woodland trail leads westward from the carpark. The best views of the reed marsh, however, are obtained by climbing onto the vantage point of Knockadoon Hill.

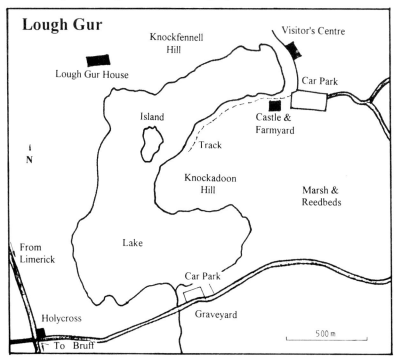

Lough Gur

Knockfennell Hill

Visitor's Centre

Lough Gur House

Car Park

Island

Castle & Farmyard

N

Track

Knockadoon Hill

Marsh & Reedbeds

From Limerick

Lake

Car Park

Holycross

Graveyard

500 m

To Bruff

Curraghchase

Curraghchase is situated 5km east of Askeaton on the main Limerick road (N69). Formerly the home of the poet Aubrey de Vere, it is now a State-owned Forest Park, with a number of visitor facilities which are open to the public throughout the year. The Forest Park is best known for its arboretum, which contains a variety of non-native trees and shrubs. There is also a small artifical lake near the manor.

Winter Flocks of finches and thrushes are plentiful, and the more unusual species seen at Curraghchase include Brambling and Hawfinch. It was for the latter species that Curraghchase became known in birdwatching circles when during a Hawfinch invasion in 1988, over 100 of these rare European thick-billed finches overwintered, feeding on the seeds of the numerous hornbeam trees.

Summer At Curraghchase Jay, Blackcap, Siskin and possibly Garden Warbler breed in the wooded area and small numbers of the common wildfowl species breed on the lake.

Birdwatching An early morning walk is recommended to see the birds of the area at their best. In the afternoons and on weekends when many people visit Curraghchase, some bird species become more wary. This is an excellent area to begin to get familiar with our common woodland birds.

Clare and Limerick ringing report

Phil Brennan
The Crag, Stonehall, Newmarket-on-Fergus, Co. Clare

Introduction

A ringing report for the region was not included in the previous bird reports. This report is an attempt to rectify this situation. With the exception of the totals for birds ringed at Adare in the seventies, and those of visiting ringers whose data is not available at the time of writing, all birds ringed in the region between 1974 and 1992 are included. The work of the Shannon Wader Ringing Group (SWRG) began in 1974, and I feel it is important that its full contribution is listed.

Bird ringers in Clare and Limerick have always been somewhat of a scarce commodity. At no time have there been more than four active ringers. All, without exception, have also been heavily involved with the IWC and other ornithological concerns. Apart from the wader group effort, the ringers have tended to seek out their own target sites and species, reacting as opportunities arose. A good volume of birds recovered shows the success of the effort. The dearth of publications is our greatest self-criticism.

Initally, the woods of Adare Manor were the focus of ringing, with a series of training courses run there in the early seventies, the last being run in 1975. The very high standard of the training meant a good start for many Irish ringers and most can trace their 'genealogy' to Adare or the Dundrum courses in Co. Down. The SWRG was the first and perhaps most significant result of the courses. The brain-child of Roger Forster and Kieran O'Brien, it started with a handful of mostly visiting ringers taking on the mud-flats of the Shannon Estuary to try to work out the comings and goings of its waders. By 1975/76, about 1000 waders were being ringed each year, with Dunlin and Curlew the main target species. Another wader project, this time on Woodcock, was carried out at Dromore Wood by the Wildlife Service from 1975-80.

With Roger Forster's departure about 1980 and Ewart Jone's retiral around the same time, I was the only active ringer left here. I concentrated on the Shannon Airport Lagoon, tried to keep the wader ringing going and started training people. Terry Carruthers, Tom Tarpey, John Murphy and Tony Mee joined the ringing effort and visiting trainees were also helped towards their permits. Since the mideighties we have targeted gulls, terns, Storm Petrel, Sedge Warbler and Sand Martin, as well as sampling new marsh sites, such as Fin Lough and Lough Gur. Whimbrel is our current wader project. Since 1985, the autumn observatory-style coverage at Loop Head has added a new dimension to the ringing, and while numbers caught are not usually very high, the site provides great opportunity for training and the experience of active migration.

Selected recoveries
Picking a good selection of recoveries is not easy. The SWRG had 96 recoveries in the period, and over 50 Sedge Warblers were recovered. This is merely a sample of some of the most interesting.

Conventions used
Controlled means caught by another ringer.
Recovered means found dead, or shot, or injured, etc.

Storm Petrel *Hydrobates pelagicus*
2322913	26.07. 85	Hag's Head, Clare	
controlled	28.07.85	Copeland Island, Down.	(2 days)
2314736	22.07.89	Loop Head, Clare	
controlled	11.08.90	Girdleness, Grampian, **Scotland**.	

Teal *Anas crecca*
M600081	27.07.81	Severn Isl., Murmansk, **USSR**.
shot	13.12.81	Islandavanna, Fergus Estuary, Clare.

Hen Harrier *Cirus cyaneus*
FV10438	04.07.80	Islay, **Scotland**
shot	01.11.83	Bunratty, Clare.

Kestrel *Falco tinninculus*
E90742	04.06.83	**Luxembourg**
found dead	04.08.83	Shannon Airport, Clare.

Golden Plover *Pluvialis apricaria*
CK83805	27.09.76	Greenish Island, Shannon Estuary	
found dead	04.07.83	Fjoll, **Iceland**.	(1546km)

Lapwing *Vanellus vanellus*
DS76401	25.01.74	Greenish Island, Shannon Est., Limerick	
shot	16.11.83	Pervomaysk, Nikolayev, **USSR**.	(2862km)
DR60559	02.02.79	nr. Aughinish Isl., Shannon Est., Limerick	
shot	08.11.83	Znanenka, Tambov, **USSR**	(3409km)

Sanderling *Calidris alba*
A colour-ringed bird from England was seen at Quilty, and a bird from Iceland was recorded at Kilkee.

Dunlin *Calidris alpina*

| NS56964 | 01.09.84 | Shannon Airport Lagoon, Clare | |
| controlled | 14.06.86 | Svalbardshr, N-Thing., **Iceland** | |

| SA747545 | 18.04.85 | Iwik, Baie Du Levrier, **Mauretania** | |
| controlled | 19.05.87 | Shannon Airport Lagoon, Clare | (3704km) |

| SUM323166 | 08.08.80 | Great Ainov Isl., Murmansk, **USSR** | |
| controlled | 22.01.81 | Tullyvarraga Marsh, Shannon | (2832km) |

A total of 79 Dunlin has been recovered as a result of SWRG work. The locations and the number of bird are:

England	21	France	4	Morocco	2
Sweden	11	Denmark	3	Iceland	1
Norway	10	Poland	2	Russia	1
Germany	7	Scotland	2	Mauretania	1
Finland	6	Portugal	2	Wales	1
Netherlands	5				

Woodcock *Scolopax rustica*

| EH34298 | 16.01.80 | Dromore Wood, Clare. |
| shot | 27.04.80 | **Estonia** |

Whimbrel *Numenius phaeopus*

| ER58012 | 28.04.91 | nr. Charleville, Limerick |
| shot | 11.08.92 | Loon-Plage, Nord, **France** |

Curlew *Numenius arquata*

| FS94622 | 14.08.75 | nr. Greenish Isl., Shannon Est., Limerick |
| Killed by raptor | 03.05.76 | Pirttikyla, Vaasa, **Finland** |

Redshank *Tringa erythropus*

| DR11113 | 03.10.89 | Shannon Airport Lagoon, Clare |
| found dead | 10.08.91 | Slugan Dubh, Mull, Strathclyde, **Scotland** |

Two other Shannon-ringed Redshank were controlled in South Uist, Scotland and Iceland, respectively. Both were nesting.

Wood Sandpiper *Tringa glareola*

| CK83869 | 23.08.74 | nr. Aughinish Isl., Shannon Est., Limerick |
| controlled | 10.08.82 | Doel, Oest- Vlaanderen, **Belgium**. |

An extraordinary recovery. This bird is the only record of its species in the estuary in modern times, and one of only two ringed by SWRG.

Sand Martin *Riparia riparia*
F655278	15.06.90	Moorestown, Kilfinane, Limerick.
controlled	16.03.91	Parc National Du Djoudj, **Senegal** (4054km)
controlled	20.07.91	Moorestown, Kilfinane, Limerick.

Six other Sand Martin from the same colony were controlled in Senegal.

Swallow *Hirundo rustica* (pullus)
F420567	10.06.89	Ballyorgan, Kilfinane, Limerick
found	23.04.90	El Hajeb, Meknes, **Morocco**.

Redwing *Turdus iliacus* (pullus)
P354866	09.06.83	Kirkkonummi, Uusimaa, Finland
killed by cat	20.01.85	Ennis, Clare

Sedge Warbler *Acrocephalus schoenobaenus*
H057689	06.08.90	Shannon Airport Lagoon, Clare.
controlled	09.08.90	Kergelan, Finistere, **France**. (3 days)

	(pullus)	
E549804	22.06.87	Larne, Antrim
controlled	06.08.87	Shannon Airport Lagoon, Clare.

H057615	01.08.90	Shannon Airport Lagoon, Clare.
retrapped	08.08.90	Shannon Airport Lagoon, Clare.
controlled	13.08.90	Wick, Dorset, **England** (4.5 days)
controlled	23.05.91	The Calf, **Isle of Man**

H057659	03.08.90	Shannon Airport Lagoon, Clare
controlled	13.12.91	Djoudj, **Senegal**
controlled	21.01.93	Djoudj, **Senegal**

A single Sedge Warbler from Fenloe, as well as three other Shannon-ringed Sedge Warblers were caught in Senegal. One of the Shannon birds was there as early as 25th October. Over 50 birds from the area have been found abroad.

Willow Warbler *Phylloscopus collybita*
893775   22.07.87   Red Island, Lough Derg, Clare.
controlled   06.08.87   Icklesham, Sussex, **England**.

Another Willow Warbler from the same small catch was caught in Jersey on 10.09.88.

Chaffinch *Fringilla coelebs*
F767365   25.03.90   Moorsholm, Cleveland, **England**.
controlled   04.11.90   Kilbaha, Loop Head, Clare.

Siskin *Carduelis spinus*
E097195   01.04.88   Castletroy, Limerick.
controlled   04.03.90   Sevenoaks, Kent, **England**.

Another Siskin from the same garden was controlled in Cleveland, **England**.

Whinchat *(Tom Tarpey)*

Ringing totals for Clare and Limeick 1974-1992

Species	Pull.	F.g.	Total	Species	Pull.	F.g.	Total
Little Grebe		3	3	Cuckoo		3	3
Fulmar		2	2	Short-eared Owl	3		3
Storm Petrel		1156	1156	Swift		22	22
Leach's Petrel		1	1	Kingfisher		10	10
Cormorant		1	1	Skylark		65	65
Shag		1	1	Sand Martin		2137	2137
Grey Heron		1.	1	Swallow	357	2132	2489
Mute Swan		7	7	House Martin		104	104
Wigeon		15	15	Tree Pipit		1	1
Teal		25	25	Meadow Pipit	12	361	372
Mallard		41	41	Rock Pipit		133	133
Pochard		1	1	Grey Wagtail	8	34	42
Hen Harrier	26		26	Pied Wagtail	3	201	204
Sparrowhawk	4	13	17	Dipper	14	27	41
Kestrel	3	6	9	Wren		397	397
Water Rail		75	75	Dunnock		328	328
Moorhen		30	30	Robin	25	920	945
Coot		2	2	Bluethroat		1	1
Oystercatcher		8	8	Black Redstart		8	8
Ringed Plover	8	22	30	Redstart		4	4
Golden Plover		9	9	Whinchat		2	2
Grey Plover		1	1	Stonechat	24	206	230
Lapwing	1	49	50	Wheatear		14	14
Knot		55	55	Gray-cheeked Thrush		1	1
Sanderling		2	2	Blackbird	36	955	991
Little Stint		1	1	Fieldfare		79	79
Curlew Sandpiper		11	11	Song Thrush	20	328	348
Purple Sandpiper		7	7	Redwing		199	199
Dunlin		4283	4283	Mistle Thrush	18	25	43
Ruff		3	3	Grasshopper Warbler		104	104
Snipe		14	75	Savi's Warbler		1	1
Woodcock		183	183	Sedge Warbler	3	6747	6750
Black-tailed Godwit		19	19	Reed Warbler		7	7
Whimbrel		186	186	Barred Warbler		1	1
Curlew		747	747	Lesser Whitethroat		1	1
Redshank		640	640	Whitethroat		48	48
Spotted Redshank		2	2	Garden Warbler		24	24
Greenshank		4	4	Blackcap		173	173
Green Sandpiper		6	6	Yellow-browed Warbler		8	8
Wood Sandpiper		2	2	Wood Warbler		1	1
Common Sandpiper		25	25	Chiffchaff		323	323
Turnstone		45	45	Willow Warbler	14	1595	1609
Laughing Gull		1	1	Goldcrest		594	594
Black-headed Gull	2122	23	2145	Firecrest		2	2
Common Gull	2		2	Spotted Flycatcher	26	11	37
Lesser Bl.-backed Gull	2	2	4	Red-breasted Flycatcher		2	2
Herring Gull	12		12	Pied Flycatcher		7	7
Great Bl.-backed Gull	6		6	Long-tailed Tit		265	265
Sandwich Tern	37		37	Coal Tit		414	414
Common Tern	222		222	Blue Tit	38	1336	1374
Woodpigeon	2	12	14	Great Tit	10	641	651
Collared Dove		13	13	Treecreeper	3	105	108

Species	Pull.	F.g.	Total	Species	Pull.	F.g.	Total
Jay		2	2	Goldfinch		34	34
Magpie	6	41	47	Siskin		77	77
Jackdaw	20	54	74	Linnet	9	196	206
Rook		45	45	Redpoll		434	434
Hooded Crow	4	5	9	Crossbill		2	2
Raven		1	1	Bullfinch		263	263
Starling		469	469	Hawfinch		2	2
House Sparrow		317	317	Yellow-rumped Warbler		1	1
Tree Sparrow		20	20	Yellowhammer		215	215
Chaffinch	4	2842	2846	Rustic Bunting		1	1
Brambling		10	10	Little Bunting		1	1
Greenfinch	5	2723	2728	Reed Bunting	2	927	929

	Pull.	F.g.	Total
Grand Total	3111	37546	40717

Conventions used in table: Pull. = pullus (young in nest); F.g. = Full grown bird.

Acknowledgements

My thanks to all the ringers mentioned in the article for their valuable contribution to the ringing effort. Thanks also to all the visiting ringers and helpers; to the land-owners who have allowed us to ring on their property and who often welcomed us into their homes; to our sponsors Aer Rianta, Shannon, De Beers Shannon and the local IWC branches.

References

Brennan, P. & Jones, E. (1982) *Birds of North Munster*. I.W.C. North Munster

Cramp, S. and Simmons, K.E.L. (eds.) (1982) *The Birds of the Western Palearctic, Vol. 3*. University Press, Oxford.

Dymond, J.N., Fraser, P.A. & Gantlet, S.J.M. (1989) *Rare Birds in Britain and Ireland*. T. & A.D. Poyser, Calton.

Forsyth, I. (1981 - 1992) Irish ringing reports, 1980 to 1991. *Irish Birds*, Vols. 2 to 4.

Gibbons, D.W., Reid, J.A. & Chapman, R.A. (1993) *The New Atlas of Breeding Birds in Britain and Ireland: 1988-91*. T & A.D. Poyser, Calton.

Green,M., Knight, A., Cartmel, S. & Thomas, D. (1988) The status of wintering waders on the non-estuarine west coast of Ireland. *Irish Birds*, 3: 569-574.

Harrison, C. (1982) *An Atlas of the Birds of the Western Palaearctic*. Collins, London.

Hutchinson, C.D. (1989) *Birds in Ireland*. T. & A.D. Poyser, Calton.

Kirby, J., Cartmel, S. & Green, M. (1991) Distribution and habitat preferences of waders wintering on the non-estuarine west coast of Ireland. *Irish Birds*, 4:317-334.

Lack, P. (1986) *The Atlas of Wintering Birds in Britain and Ireland*. T. & A.D. Poyser, Calton.

Lysaght, L.S. (1986) The Birds of Westfields, Limerick City. *Irish Birds*, 3: 255-266.

Lysaght, L.S. (1989) Breeding bird populations of farmland in mid-west Ireland in 1987. *Bird Study*, 36:91-98.

Macdonald, R.A. (1987) The breeding population and distribution of the Cormorant in Ireland. *Irish Birds*, 3: 405-416.

Mayes, E. & Stowe, T. (1989) The status and distribution of the Corncrake in Ireland, 1988. *Irish Birds*,4: 1-12.

Mead, C. (1974) *Bird Ringing*. British Trust for Ornithology, Tring.

Merne, O.J. & Murphy, C.W. (1986) Whooper Swans in Ireland, January 1986. *Irish Birds*, 3: 199-206.

O'Sullivan, O. & Smiddy, P. (1988) Thirty-fifth Irish Bird Report, 1987. *Irish Birds*, 3: 609-648

Preston, K., Smiddy, P. et al (1982-1991) Irish Bird Report [Annual from 1982-1991]. *Irish Birds*,

Reynolds, J.W. (1990) The breeding gulls and terns of Lough Derg. *Irish Birds*, 4:217-226.

Ruttledge, R.F. (1987) The breeding distribution of the Common Scoter in Ireland. *Irish Birds*, 3: 417-426.

Sheppard, R. (1993) *Ireland's Wetland Wealth*. I.W.C. report.

Stapleton, L. (ed.) (1975) *Birds of Clare and Limerick*. I.W.C. North Munster Branch.

Whilde, A. (1985) The 1984 All Ireland Tern Survey. *Irish Birds*, 3: 1-32.

Index to species list